6th International Conference on Computer Science and Computational Mathematics (ICCSCM 2017)

Journal of Physics: Conference Series Volume 892

Langkawi, Malaysia
4 – 5 May 2017

ISBN: 978-1-5108-4866-5
ISSN: 1742-6588

Printed from e-media with permission by:

Curran Associates, Inc.
57 Morehouse Lane
Red Hook, NY 12571

proceedings.com

Some format issues inherent in the e-media version may also appear in this print version.

Printed by Curran Associates, Inc. (2017)

For permission requests, please contact the Institute of Physics
at the address below.

Institute of Physics
Dirac House, Temple Back
Bristol BS1 6BE UK

Phone: 44 1 17 929 7481
Fax: 44 1 17 920 0979

techtracking@iop.org

Additional copies of this publication are available from:

Curran Associates, Inc.
57 Morehouse Lane
Red Hook, NY 12571 USA
Phone: 845-758-0400
Fax: 845-758-2633
Email: curran@proceedings.com
Web: www.proceedings.com

Table of contents

Volume 892

The 6th International Conference on Computer Science and Computational Mathematics (ICCSCM 2017)

4–5 May 2017, Langkawi, Malaysia

Accepted papers received: 5 September 2017
Published online: 21 September 2017

Preface

Papers

IOP Conf. Series: Journal of Physics: Conf. Series **892** (2017) 011001 doi:10.1088/1742-6596/892/1/011001

The ICCSCM 2017 (The 6th International Conference on Computer Science and Computational Mathematics) has aimed to provide a platform to discuss computer science and mathematics related issues including Algebraic Geometry, Algebraic Topology, Approximation Theory, Calculus of Variations, Category Theory; Homological Algebra, Coding Theory, Combinatorics, Control Theory, Cryptology, Geometry, Difference and Functional Equations, Discrete Mathematics, Dynamical Systems and Ergodic Theory, Field Theory and Polynomials, Fluid Mechanics and Solid Mechanics, Fourier Analysis, Functional Analysis, Functions of a Complex Variable, Fuzzy Mathematics, Game Theory, General Algebraic Systems, Graph Theory, Group Theory and Generalizations, Image Processing, Signal Processing and Tomography, Information Fusion, Integral Equations, Lattices, Algebraic Structures, Linear and Multilinear Algebra; Matrix Theory, Mathematical Biology and Other Natural Sciences, Mathematical Economics and Financial Mathematics, Mathematical Physics, Measure Theory and Integration, Neutrosophic Mathematics, Number Theory, Numerical Analysis, Operations Research, Optimization, Operator Theory, Ordinary and Partial Differential Equations, Potential Theory, Real Functions, Rings and Algebras, Statistical Mechanics, Structure Of Matter, Topological Groups, Wavelets and Wavelet Transforms, 3G/4G Network Evolutions, Ad-Hoc, Mobile, Wireless Networks and Mobile Computing, Agent Computing & Multi-Agents Systems, All topics related Image/Signal Processing, Any topics related Computer Networks, Any topics related ISO SC-27 and SC- 17 standards, Any topics related PKI(Public Key Intrastructures), Artifial Intelligences(A.I.) & Pattern/Image Recognitions, Authentication/Authorization Issues, Biometric authentication and algorithms, CDMA/GSM Communication Protocols, Combinatorics, Graph Theory, and Analysis of Algorithms, Cryptography and Foundation of Computer Security, Data Base(D.B.) Management & Information Retrievals, Data Mining, Web Image Mining, & Applications, Defining Spectrum Rights and Open Spectrum Solutions, E-Comerce, Ubiquitous, RFID, Applications, Fingerprint /Hand/Biometrics Recognitions and Technologies, Foundations of High-performance Computing, IC-card Security, OTP, and Key Management Issues, IDS/Firewall, Anti-Spam mail, Anti-virus issues, Mobile Computing for E-Commerce, Network Security Applications, Neural Networks and Biomedical Simulations, Quality of Services and Communication Protocols, Quantum Computing, Coding, and Error Controls, Satellite and Optical Communication Systems, Theory of Parallel Processing and Distributed Computing, Virtual Visions, 3-D Object Retrievals, & Virtual Simulations, Wireless Access Security, etc. The success of ICCSCM 2017 is reflected in the received papers from authors around the world from several countries which allows a highly multinational and multicultural idea and experience exchange.

The accepted papers of ICCSCM 2017 are published in this Book. Please check http://www.iccscm.com for further news.

A conference such as ICCSCM 2017 can only become successful using a team effort, so herewith we want to thank the International Technical Committee and the Reviewers for their efforts in the review process as well as their valuable advices. We are thankful to all those who contributed to the success of ICCSCM 2017 .

The Secretary

Peer review statement

All papers published in this volume of *Journal of Physics: Conference Series* have been peer reviewed through processes administered by the proceedings Editors. Reviews were conducted by expert referees to the professional and scientific standards expected of a proceedings journal published by IOP Publishing.

IOP Conf. Series: Journal of Physics: Conf. Series **892** (2017) 012001 doi:10.1088/1742-6596/892/1/012001

Utilizing Centralized Diamond Architecture for Searching Algorithms

Masumeh Damrudi, Kamal Jadidy Aval

Computer Science Department, Islamic Azad University, Firoozkooh Branch, 3981838381, Iran

m.damrudi@yahoo.com, k.jadidy@yahoo.com

Abstract. Searching data elements and information in a fraction of time is still an important task for many areas of works. The importance of finding the best answer leads to use different algorithms which consume time. To improve the speed of searching information within mass information, employing parallel processing is inevitable. A search algorithm with constant time(CS) is proposed on centralized diamond architecture. We have optimized this algorithm to find the location and the number of occurrence of data, if there exists any.

1. Introduction
Finding the appropriate answer in the search algorithms is a critical work in some areas of study. In computer science and mathematics, a search algorithm receives an input and after evaluating the possible solutions, returns the answer. Various searching algorithms are presented by different researchers which are employed in applications. Some of these applications include expert systems, database systems, game systems, and robot control systems [1].

Some of the search algorithms are considered in [2]. A parallel search algorithm on sorted matrices with $O(1loglogn)$ execution time using $O(n/loglog\ n)$ processors on a CREW PRAM is presented in [3]. Their selection is considered to be an n×m sorted matrix. Two parallel search algorithms with different complexities in different ways are explained on [4]. A parallel search algorithm on tree is presented with $O(log2(n+1))$ time complexity in [5].

We issue an improved algorithm on the new searching algorithm on Centralized Diamond architecture presented in [6]. Diamond architecture which is heterogeneous was proposed in [7]. This architecture has $(7/4)n$ processor elements. NOD and ENOD sorting algorithms are using this architecture [7, 8]. The Centralized Diamond issued in [6], is a variation of Diamond exploiting one more node in the center. This improved architecture leads to a better result in the design of algorithm which the search algorithms are presented on centralized diamond architecture.

The first search algorithm is the improved of CS algorithm [6] which give is getting the key value location as the result of search too. The second search algorithm count the number of searched value as the result of algorithm.

2. Improved of CS Search on Centralized Diamond Architecture
The main CS search algorithm and the Centralized Diamond architecture are completely explained in [6]. In this paper, assume n is the number of data elements and N is the number of processor elements. The Centralized Diamond architecture for $n=16$ and $N=29$ is shown in Figure 1.

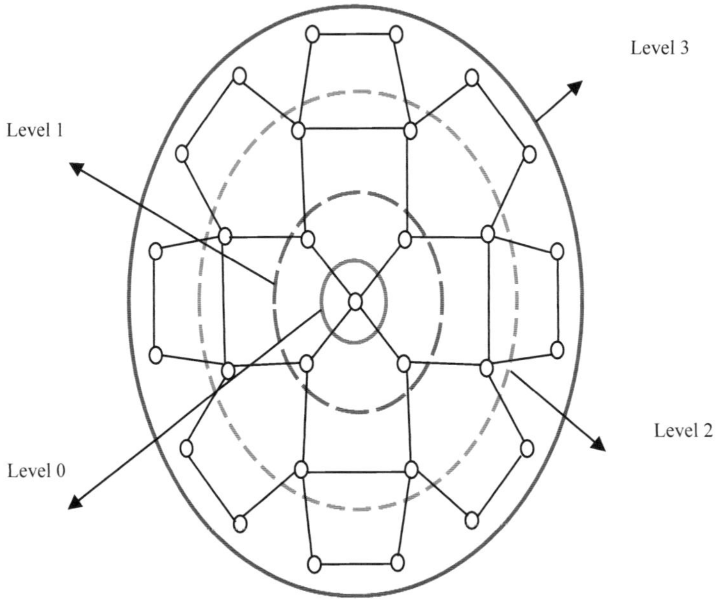

Figure 1. Centralized Diamond architecture

In the first place, all data elements are in the third level as inputs. The aim is to search the key value of *k* among data elements as well as its location. It means that if the data element exists, which processor element has it. It is assumed that the data elements are distinct. Step 1: At first, k is entered to all nodes in the third layer of the architecture at the same time. All nodes compare their data elements with the received value k. If these two data were equal in the node, the node sends the PE number to its parent which means the data exists as shown in Figure 2. Otherwise the node sends 0 to the parent.

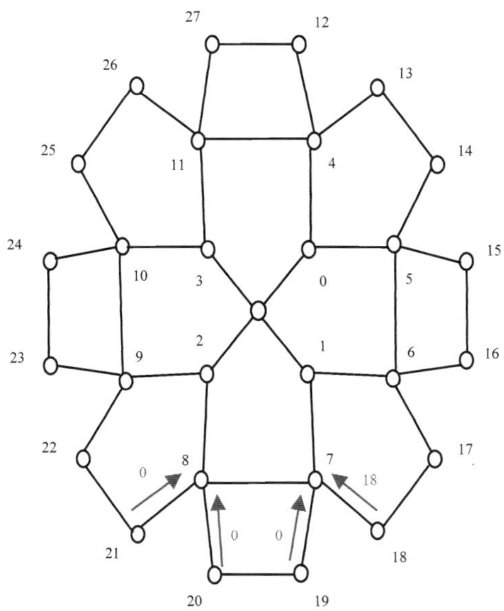

Figure 2. Operations in the third layer

In this figure, node 18 has the *k* and sends PE number which is 18 to its parent. Step 2: At the second level, each processor performs OR operation with the received values which are 0 or a PE number. The result of OR operation will be sent to its parent in the first level. Figure 3 shows this operation.

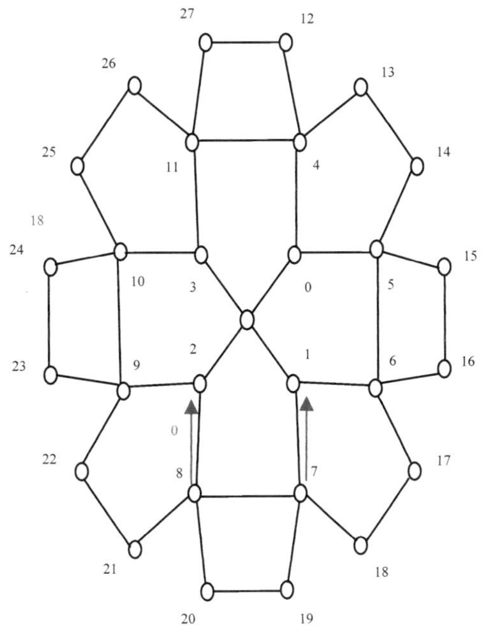

Figure 3. Operations in the second layer

Step 3: At the first level, each processor element makes OR operation with the received values. The result of OR operation will be sent to its parent which is the central node in the zero level. The received values are either zero or number of a PE that has the *k* value. Figure 4 shows this operation.

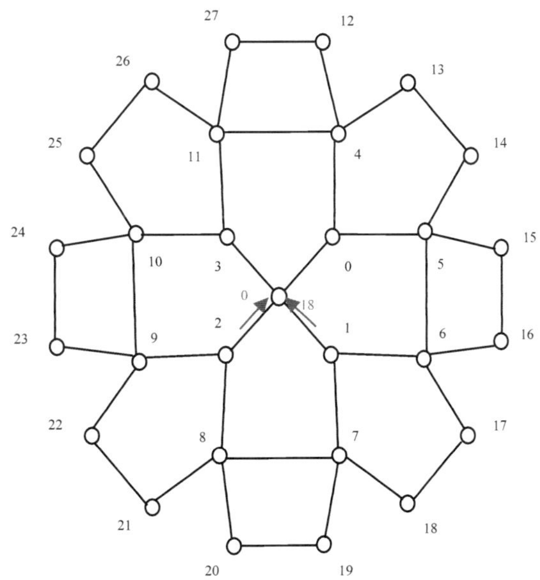

Figure 4. Operations in the first layer

Step 4: Level number zero receives the results from first level. These data are the number of a PE or zeros. The OR operation on the received values present a number. If this number is 0, it means the key *k*, doesn't exist and if the number becomes greater than 0, it means that data exists and the number is the PE number which has the key.

3. Counting of Searching Value on CS Search

In some cases, it is important that how many of the searching value exists. This algorithm is called counting. The algorithm is explained via an example. In this example, it is considered that PEs 18 and 21 have the *k* value. At the first place, all data elements are in the third level as inputs. Step 1: At first, *k* is entered to all nodes in the third layer of the architecture at the same time. All nodes compare their data elements with the received value *k*. If these two data were equal in the node, the node sends 1 to its parent which means the data exists. Otherwise the node sends 0 to the parent. The results are shown in Figure 5. It is considered that the PEs with number 18 and 21 have the key value.

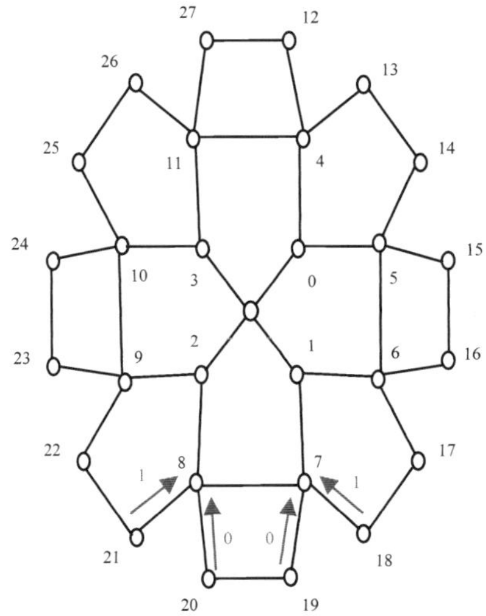

Figure 5. Operations in the third layer

Step 2: At the second level, each processor calculates the sum of received values which are 0 or 1. The result of addition will be sent to its parent in the first level. Figure 6 shows this operation.

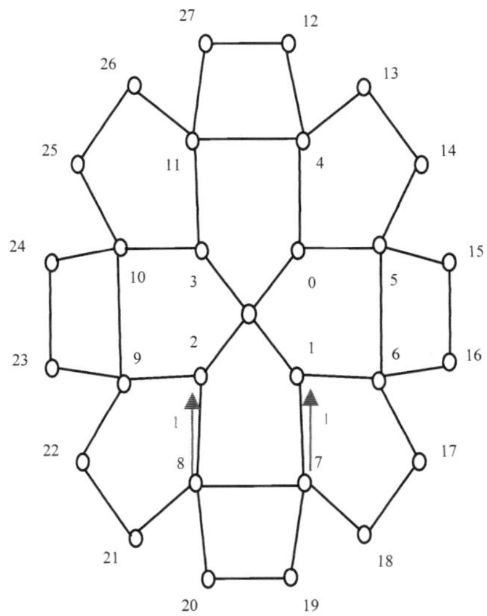

Figure 6. Operations in the second layer

Step 3: At the first level, each processor element adds the received values. The result of addition will be sent to its parent which is the central node in the zero level. The received values are either zero or a number greater than zero. Figure 7 shows this operation.

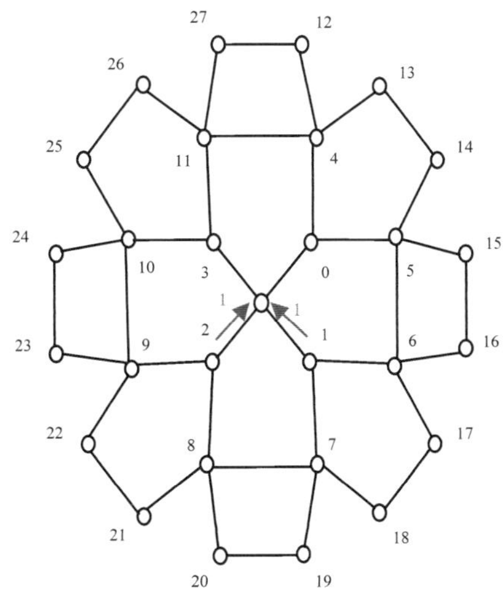

Figure 7. Operations in the first layer

Step 4: Level number zero receives the results from first level. The addition of the received values presents a number. If this number is 0, it means the key *k*, doesn't exist and if the number becomes greater than 0, it means that data exists and the number shows how many of data k exist.

4. Analysis and Cost

Total execution time for both searching algorithms is constant. In the improved of CS search algorithm, Step one has just one comparison, which is *O(1)*. In step two, two received data elements perform OR operation which take place in *O(1)*. The third step is just like the second step which means it needs *O(1)* execution time. The last step requires two OR operations. This step requires *O(2)*.

The total execution time is constant $t_n=O(5)=O(1)$, therefore the total cost is the number of processor elements $c_n=O(N)$.

The execution time for the second algorithm (counting algorithm) is the same as the improved CS algorithm. The difference is the type of operations. The operations in the second algorithm is addition while in the first algorithm is OR operation.

Considering the first algorithm, in following example assume that we are looking for $k= 5$ among data elements. It's seen that having $N=29$ and $n=16$ is like four threes are connected to each other. The number of trees differs by the number of data elements and processor elements as it's considered in Figure 8.

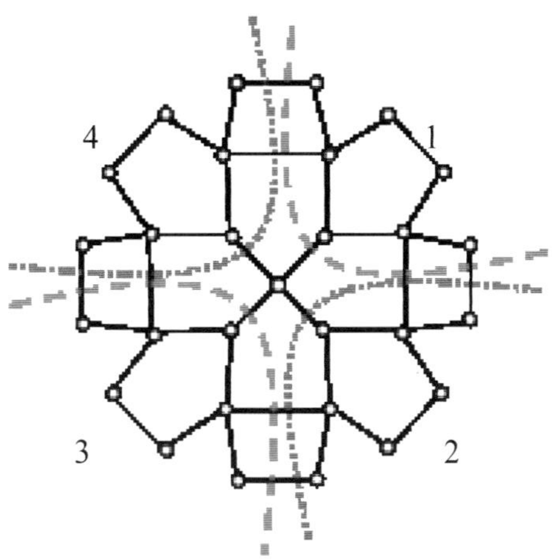

Figure 8. Division of Centralized Diamond

All these four parts work in parallel during the algorithms. The steps from one to three are shown in Figure 9 for improved CS algorithm.

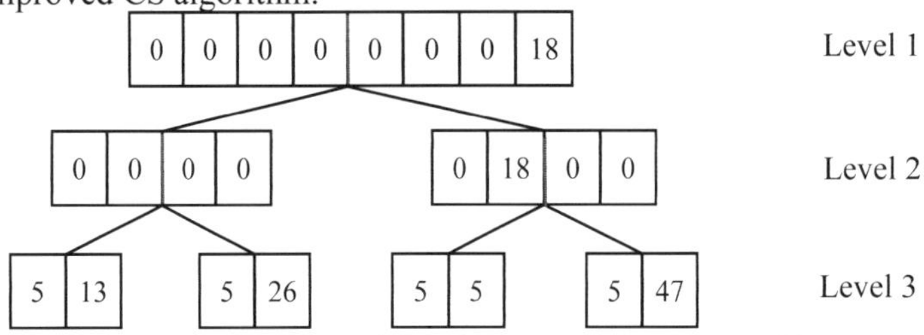

Figure 9. The procedure of algorithm from step 1 to step 3 of improved CS algorithm

In step one, the architecture from the left side, 13 is not equal to the key 5, so sends 0 to level two and also 26 is not equal too, then it sends 0 to level 2 as shown in the figure. The value 5 is equal to the key 5, then the value 18 which is the number of PE is sent to level 2. The squares show the number of registers of each PEs in each level.

In level two, in the left side, two received zeros do OR operation, and the result which is 0 is sent to the first level. In the right side, the OR operation of zero and 18 is 18. The result is sent to the level one.

In level 1, one of two received values is 18 then 18 as the results of OR is sent to level 0 because this is the result of third section in Figure 8. The other sections 1, 2 and 4 do not have the key while the third section has it. It is because the data are assumed to be distinct. You see the procedure of step 4 in Figure 10.

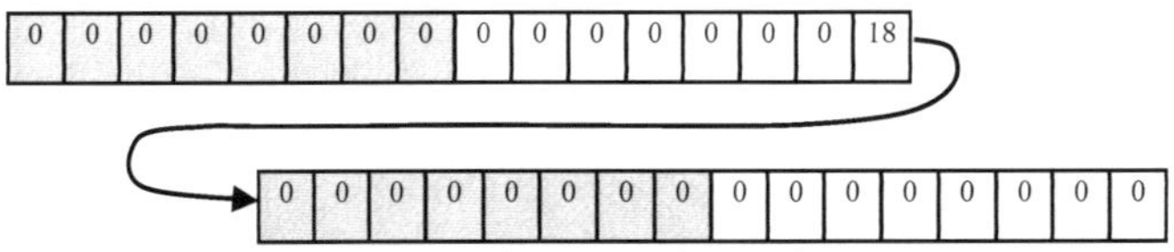

Figure 10. The procedure of step 4

In the last step, the centralized node does OR operation among the four received data and the final result is 18.

The second algorithm (counting algorithm) is the same as the first algorithm. The difference is the operation. Figure 11 shows the steps of this algorithm for the following example.

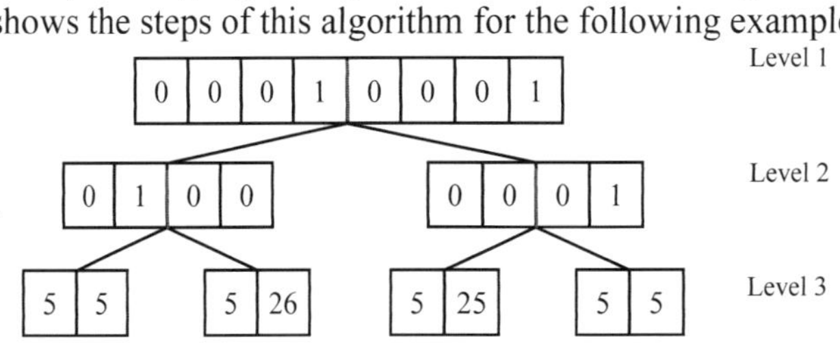

Figure 11. The procedure of algorithm from step 1 to step 3 of counting algorithmt

In step one, the architecture from the left side, 5 is equal to the key 5, so sends 1 to level two and also 26 is not equal to key 5, then it sends 0 to level 2 as shown in the figure. The value 25 is not equal to the key 5, then the value 1 is sent to level 2.

In the level two, in the left side, two received values do addition operation, the result which is 1 is sent to the first level. In the right side, the addition operation of zero and 1 is 1, and the result is sent to the level one.

In level 1, the two received values are 1 then, the sum of these two values which is 2, as the results of addition is sent to level zero. This is the result of third section in Figure 8 for counting algorithm. The results of other sections are calculated and sent to level zero like the third section. The procedure of step 4 is shown in Figure 12.

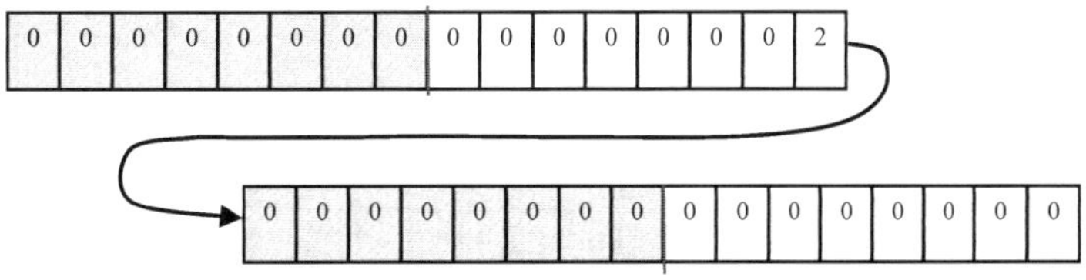

Figure 12. The procedure of step 4

In the last step, the centralized node adds the results from four sections and the final result in this example is 2.

5. Conclusion

We proposed the improved of our previous search algorithm (CS) with constant execution time which gives the place of the search value as the result. The proposed algorithm has a suitable order using the centralized diamond architecture where the total cost is . It is assumed that the data is distinct in this algorithm.

The second proposed algorithm using mentioned architecture is counting which shows how many of the searched data exists. This algorithm needs constant execution time too.

References

[1] M. Damrudi, K. Jadidy Aval, 2011, Searching data using centralized diamond architecture, *Journal of Communication and Computer*, 8(9) , 807-811.
[2] D. Eppstein, Z. Galil, 1988, Parallel algorithmic techniques for combinational computation, *Annual review of computer science*, 3(1), 233-283.
[3] R. Sarnath, X. He, 1992, Efficient parallel algorithms for selection and searching on sorted matrices, In *Proceedings of Sixth International Parallel Processing Symposium*, Washington, DC, USA, 1992, 108-111.
[4] K.T. Yar, N.S.S Aung. 2009, The optimal running time from computational models by comparing two parallel searching procedures, In *IEEE International Workshop on Open-source Software for Scientific Computation (OSSC)*, china, 2009, 73-78.
[5] K. A. Berman, J. L. Paul, 2005, *Algorithms: sequential, parallel, and distributed*, Course Technology Ptr, USA.
[6] M. Damrudi, Masumeh, K. Jadidy Aval, 2011, Searching data using centralized diamond architecture, *Journal of Communication and Computer*, 8(9), 807-811.
[7] M. Damrudi, K. Jadidy Aval, 2009, Diamond architecture with NOD sorting, *IEEE Youth Conference In Information, Computing and Telecommunication*, 2009, 431-434.
[8] M. Damrudi, K. Jadidy Aval, 2010, A new parallel sorting on Diamond architecture," In *Proc. 4th Conf. Europ. Comput. Conf.*, Univ. Politechnica of Bucharest, Bucharest, Romania, 2010, 284-287.

An integrated study of surface roughness in EDM process using regression analysis and GSO algorithm

Nurezayana Zainal[1], Azlan Mohd Zain[1], Safian Sharif[2], Haza Nuzly Abdull Hamed[1] and Suhaila Mohamad Yusuf[1]

[1]Soft Computing Research Group, Faculty of Computing, Universiti Teknologi Malaysia, 81310 Skudai, Johor, Malaysia

[2] Faculty of Mechanical Engineering, Universiti Teknologi Malaysia, 81310 Skudai, Johor, Malaysia

azlanmz@utm.my

Abstract. The aim of this study is to develop an integrated study of surface roughness (Ra) in the die-sinking electrical discharge machining (EDM) process of Ti-6AL-4V titanium alloy with positive polarity of copper-tungsten (Cu-W) electrode. Regression analysis and glowworm swarm optimization (GSO) algorithm were considered for modelling and optimization process. Pulse on time (A), pulse off time (B), peak current (C) and servo voltage (D) were selected as the machining parameters with various levels. The experiments have been conducted based on the two levels of full factorial design with an added center point design of experiments (DOE). Moreover, mathematical models with linear and 2 factor interaction (2FI) effects of the parameters chosen were developed. The validity test of the fit and the adequacy of the developed mathematical models have been carried out by using analysis of variance (ANOVA) and F-test. The statistical analysis showed that the 2FI model outperformed with the most minimal value of Ra compared to the linear model and experimental result.

1. Introduction

Managing quality has been a key determinant in an organizations' in order to get lower total costs as well. The high quality of finished products includes of high surface finish, less tool wear and high production rate, but in term of economy machining is lower cost [1]. Therefore, the selection of optimum process parameters plays a significant role. The success of the manufacturing process is determined upon by the selection of appropriate process parameters [2]. The purpose of the optimization process is to look for a solution which is as close as possible to the target and as robust as possible [3,4,5]. The first step for process parameter optimization is developing a mathematical model in order to have a better understanding the principles leading the machining process [6]. Basically, a model has to be developed with the aid of mathematical equations and requires input data (control parameters).

In manufacturing industries, machining process is one of the most important and widely being used compared to forming, casting and joining processes [7,8]. Basically, there are two types of machining processes which are traditional (milling, turning, grinding, drilling and others) and modern machining (abrasive water jet, electrical discharge machining, electrochemical machining and others) [9].

Nowadays, electrical discharge machining (EDM) is one of the widely used machining processes due to its capability of machining geometrically complex or hard material component which need precise and high accuracy [10].

Moreover, EDM is an alternative method of serial or batch production of difficult-to-cut parts when it is not possible to create by using conventional machining methods [11]. In this study, die-sinking EDM process has been selected for conducting experiments. Basically, in a die-sinking EDM, material is removed by controlling erosion through a series of electric sparks between the tool (electrode) and the workpiece to have the eroding effect on work-piece in order to form a replica of tool on work piece [12]. In this process an electric spark is used as the cutting tool to cut (erode) the workpiece to produce the finished part to the desired shape [13]. Thousands of electrical discharges per second are generated and discharge produces a crater by melting and vaporization [13, 14]. Some melted material is flushed away by the dielectric fluid [11]. A dielectric fluid not only flushes out the chips, but also confines the electric discharge. Thus a perfect reproduction of the shape of the tool on the workpiece is reproduced [15, 16].

Surface roughness (Ra) will play a critical role in evaluating and measuring the surface quality of the machined product. Ra greatly affects the functional attributes of products which are frictionless, wear resistance, fatigue, lubricant, light reflection, and coating. It is a measure of the technological quality of a product and it is a factor that significantly affects the manufacturing cost as well [17, 18]. Ra was known as a significant outcome in the manufacturing process and it materializes a major part in the manufacturing system. The quality of finished products was determined by how closely the finished product adheres to certain qualifications including the dimensions and surface quality. Ra value is influenced by many factors such as machining parameters, cutting phenomena, workpiece properties and cutting tool properties and its prediction and control is a demand to the researchers [9].

There are many literatures with the relevant investigations on the integrated study of surface roughness. Patel et al. [19] developed surface roughness prediction model for EDM of Al2O3/SiCw/TiC ceramic composite on discharge current, pulse on time, duty cycle and gap voltage. The model was developed based on the response surface methodology. It was observed that the optimization process by using trust-region method gives the optimal value of surface roughness (1.88 μm) at the combination setting of 3A of discharge current, 10μs of pulse on time and 0.88 of duty cycle. It can be concluded that the developed models can predict the surface roughness accurately within 95% confidence interval.

Bhattacharya et al. [20] proposed a study on surface integrity model for the EDM of M2 Die Steel. The mathematical models were developed for surface roughness, recast layer thickness, and surface crack density based on the response surface methodology (RSM) which considered peak current and pulse on duration as the machining parameters. It is observed that the peak current and pulse duration should be kept as low as possible in order to achieve the minimum value of surface roughness.

Mandal et al. [21] presented a study on modelling and optimization of machining nimonic C-263 super alloy using multicut strategy in wire electrical discharge machining (WEDM). The mathematical modelling for cutting rate, surface roughness, spark gap and wire wear ration were developed by using regression analysis, which considered pulse on time (TON), pulse off time (TOFF), servo voltage (SV) and dielectric flow rate (FR) as the machining parameters. It can be observed that all the mathematical models suggested are fairly fitted with the experimental result with a confidence level of 95%. Moreover, the minimum surface roughness value of 1.38 mm can be achieved by optimal setting of controlling parameters (TON 0.95μs, TOFF 200μs, SV 75V, and FR 8L/min). Thus, it can be concluded that the multicut strategy improves the surface roughness of the work piece.

Gopalakannan et al. [22] employed RSM approach for the prediction of material removal rate (MRR), electrode wear rate (EWR) and surface roughness (SR) of Al 7075-B4C MMC. Pulse current, gap voltage, pulse on time and pulse off time were selected as the cutting parameters. It was reported that the proposed mathematical models match the experimental values reasonably well with R^2 of 0.9838 for MRR, 0.9627 for EWR and 0.9875 for SR.

Anitha et al. [23] presented a multi-objective optimization approach by modelling MRR and SR of EDM process using artificial neural network (ANN) approach. The study considered pulse current, pulse duration, duty cycle and voltage as the input process parameters. The results show an improvement with a better productivity, a reduced material removal time and a product cost at the desired surface finish. The optimized value of the present research is found to be 51.588 mm3/min with a level of surface finish of 0.0955 μm.

Thus, this study carried out an experimental design in order to conduct a comprehensive investigation on the influence of EDM parameters towards the Ra. Furthermore, the study will apply regression analysis and glowworm swarm optimization (GSO) algorithm for modeling and optimization processes.

2. Experimentation

A number of experiments were conducted on a die-sinking electrical discharge machine, model AG40L Sodick. Ti-6Al-4V titanium alloy has been selected as a work piece material while copper-tungsten was considered as an electrode. EDM oil was used as dielectric fluid. The mechanical properties and chemical composition of the Ti-6Al-4V are illustrated as in Table 1. The Ti-6Al-4V titanium alloy was cut to a size of 100mm × 60mm × 5mm for experimental and machined using 8mm diameter of the electrodes. Machining depth was kept constant at 1.5mm for every experimental run. The machining parameters considered in the study are pulse on time (A), pulse off time (B), peak current (C) and servo voltage (D). Meanwhile, the design factors (machining parameters) with their designation, units and respective levels are listed in Table 2. The experimental setup can be seen as in Figure 1.

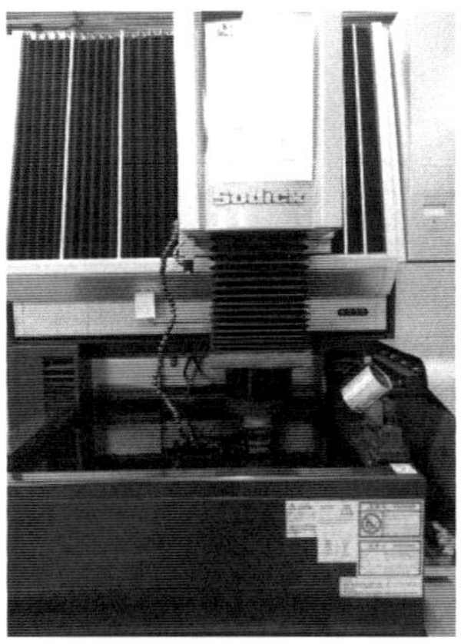

Figure 1. Experimental setup using EDM machine

Table 1. Chemical composition and mechanical properties of Ti-6Al-4V.

Mechanical composition		Mechanical properties	
Al	6.37	Tensile strength (MPa)	960–1270
V	3.89	Yield strength (MPa)	820
Fe	0.16	Elongation 5D (%)	≥8
C	0.002	Reduction in area (%)	≥25
Mo	<0_01	Density (g/cm3)	4.42
Mn	<0_01	Modulus of elasticity tension (GPa)	100–130

Si	<0_01	Hardness (Hv)	330–370
Ti	Balance	Thermal conductivity (W/mK)	7

Table 2. Factor and level for EDM parameters of titanium alloy.

Symbol	Name	Units	Level	
			Low (-1)	High (+1)
A	Pulse on time	μs	150	230
B	Pulse off time	μs	60	90
C	Peak current	A	10	12
D	Servo voltage	Volt	30	60

The experiment was run according to the full factorial design of experiment methodology with added four center points which resulted in a total number of 20 experimental runs. The arrangement of the experiment's parameters was noted as coded terms; the low level (-), center point (cp) and high level (+) as presented in Table 3.

Table 3. Two level full factorial experiments with four factors and four center points.

Run	Machining parameters			
	A (μs)	B (μs)	C (A)	SV (V)
1	-	-	-	-
2	+	-	-	-
3	-	+	-	-
4	+	+	-	-
5	-	-	+	-
6	+	-	+	`-
7	-	+	+	-
8	+	+	+	-
9	-	-	-	+
10	+	-	-	+
11	-	+	-	+
12	+	+	-	+
13	-	-	+	+
14	+	-	+	+
15	-	+	+	+
16	+	+	+	+
17	cp	cp	cp	cp
18	cp	cp	cp	cp
19	cp	cp	cp	cp
20	cp	cp	cp	cp

IOP Conf. Series: Journal of Physics: Conf. Series **892** (2017) 012002 doi:10.1088/1742-6596/892/1/012002

2.1. Experimental results

The Ra values of the machined workpiece were measured by using Mitutoyo Formtracer surface roughness tester. Before conducting the measurement, the workpiece was blown using a hair dryer to ensure no debris and dielectric fluid was present. Basically, the measurement of surface roughness needed was repeated for three times. Then, the average value of Ra was taken. The experimental run orders and their results were presented in Table 4.

Table 4. Ra values of the machining experiment.

Run	Machining parameters				Machining response
	A (μs)	B (μs)	C (A)	SV (V)	Ra (μm)
1	150	60	10	30	2.6106
2	230	60	10	30	3.7860
3	150	90	10	30	2.6137
4	230	90	10	30	3.5734
5	150	60	12	30	3.4860
6	230	60	12	30	3.4398
7	150	90	12	30	2.5694
8	230	90	12	30	2.9312
9	150	60	10	60	2.4475
10	230	60	10	60	2.3038
11	150	90	10	60	2.2949
12	230	90	10	60	2.4646
13	150	60	12	60	2.5322
14	230	60	12	60	2.5121
15	150	90	12	60	2.5067
16	230	90	12	60	2.5430
17	190	75	11	45	2.3455
18	190	75	11	45	2.6518
19	190	75	11	45	2.5493
20	190	75	11	45	2.4989

Moreover, the machining layout of work piece and electrodes are shown in Figure 2 and Figure 3, respectively.

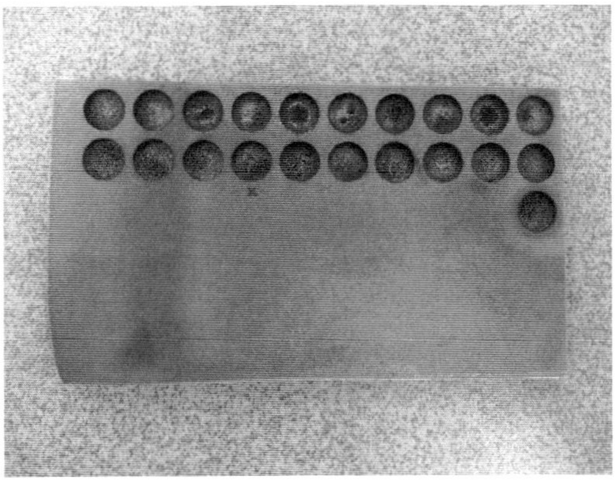

Figure 2. Ti-6AI-4L titanium alloy work piece after machining

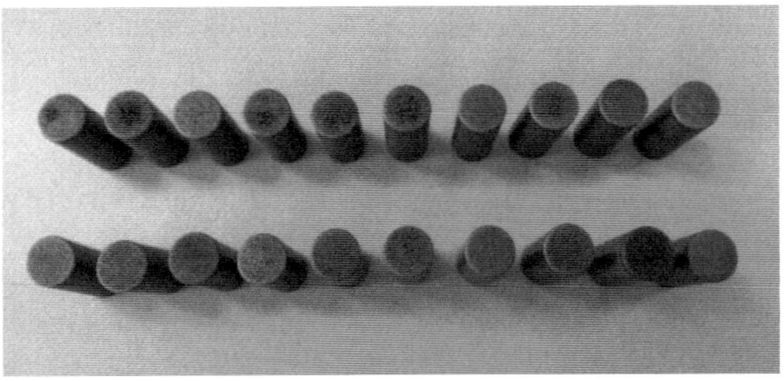

Figure 3. Copper-tungsten electrodes after machining

3. Modeling

Regression analysis is a conceptually simple method for investigating functional relationships among variables. The relationship is expressed in the form of an equation or a model connecting the response or dependent variable and one or more predictor or independent variables [22]. Regression equations are found out using software of statistical analysis called Design Expert Version 6.0.4. To gain this job, linear and 2 factor interaction (2FI) models, both were exercised to investigate the parametric effect on response. We denote the response variable by Y and the set of predictor variables by $x_1, x_2,...,x_n$, where n represents the number of independent variables. The relationship between Y and $x_1, x_2,...,x_n$ can be approximated by the regression model as in (1).

$$Y = \beta_0 + \beta_1 x_1 + \beta_2 x_2 + \beta_3 x_3 + \beta_4 x_4 + \beta_{12} x_1 x_2 + \beta_{13} x_1 x_3 + \beta_{14} x_1 x_4 + \beta_{23} x_2 x_3 + \beta_{24} x_2 x_4 + \beta_{34} x_3 x_4 \quad (1)$$

Where Y is the estimated responses (Ra), x_n are the variables for four machining parameters (A, B, C and D) and β_n are regression parameters or coefficients. The coefficient β_i are the linear terms while the coefficient β_{ij} are the interaction terms.

3.1 Regression models of surface roughness

According to the results of machining experiments given in Table 4, regression models for both linear and 2FI models are developed. The equations of the fitted model for Ra according to the four variables (x_1= A, x_2= B, x_3= C and x_4= D) as in (1) can be written as follows:

$$\mathrm{Ra} = 3.2756 + (0.0039\mathrm{x}_1) - (0.0068\mathrm{x}_2) + (0.0266\mathrm{x}_3) - (0.0225\mathrm{x}_4)$$

(2)

$$Ra = -5.1249 + 0.0422x_1 + 0.0174x_2 + 0.8116x_3 - 0.0431x_4 + (5.8552E-05x_1x_2) - 0.0029x_1x_3 - 0.0003x_1x_4 - 0.0051x_2x_3 + 0.0005x_2x_4 + 0.0031x_3x_4$$

(3)

3.2 Model adequacy test

The analysis of variance (ANOVA) and F-test were performed to justify the goodness of fit of the developed mathematical models as well as the significance of individual machining parameters. Table 5 shows the pre-ANOVA model summary statistic of Ra. It can be seen that the value of R-squared for the 2FI model is calculated as 0.835, which is much better as it is near to 1 compared to the linear model (0.608). Moreover, it indicates that the presented data fit in the developmental model. Hence, the 2FI model suggested is the most appropriate.

Table 5. Model summary statistics for Ra.

Model	R^2	Adj. R^2	Pred. R^2
Linear	0.608	0.504	0.169
[a]2FI	0.835	0.651	-0.088

[a]suggested

Table 6 and Table 7 shows the summary of the ANOVA results of the proposed models for Ra. The ANOVA table contains the degree of freedom (DF), sum of square (SS), mean of square (MS), F-value and P-value. As seen from Table 6, the developed linear model is significant with the P-value of 0.0049, which indicates that the model is significant at 95% confidence level. The suggested 2FI model also statistically significant for Ra which is 0.0163. Moreover, the MS value of the linear model (0.5975) is larger than the value of MS of residual (0.1026), thus the computed F value is 5.8265 implies that the model is significant. As presented in Table 7, it can be observed that the MS value of the 2FI model is 0.3275, which is also larger that the MS value of the residual (0.0722). Hence, the computed F value implies the model is also significant (4.5397). Table 8 and 9 both show the value of R-squared (R^2), adjusted R-squared (Adj. R^2) and predicted R-squared (Pred. R^2) statistics. R^2 value indicates the adequacy of the suggested model. The highest R^2 value, the better the model fits the experimental data, which is R^2 is always between 0 and 100%. Linear model shows that 60.8% of the variation is attributed to process parameters. 2FI model shows higher values of R^2 compared to the linear model which explains variations in the Ra to the extent of 83.5%. The suggested model with the higher of Adj. R^2 values are the better fit variables of the model. We can see that, 2FI model produced higher values of Adj. R^2 (65.1%) compared to the linear model (50.4%) which explains the addition of interaction variables in the model gets the model better fit. Pred. R^2 determines how well the model predicts responses for new observations. Larger values of Pred. R^2 indicate models of greater predictive ability. For linear model, the Pred. R^2 of 0.169 is not reasonable agreement with the Adj. R^2 of 0.504 because the difference between the adjusted and predicted R^2 is more than 0.2 which was not recommended for model to be adequate. 2FI model also a negative value of predicted R^2 (-0.088) implies that the overall mean is a better predictor of the Ra.

Table 6. ANOVA for linear model of Ra.

Source	DF	SS	MS	F value	P value
Regression	4	2.3901	0.5975	5.8265	0.0049

Residual error	15	1.5383	0.1026		
Total	19	3.9283			

Table 7. ANOVA for 2FI model of Ra.

Source	DF	SS	MS	F value	P value
Regression	10	3.2784	0.3278	4.5397	0.0163
Residual error	9	0.6499	0.0722		
Total	19	3.9283			

Table 8. ANOVA model adequacy for linear model of Ra.

R^2	0.608
Adj. R^2	0.504
Pred. R^2	0.169

Table 9. ANOVA model adequacy for 2FI model of Ra.

R^2	0.835
Adj. R^2	0.651
Pred. R^2	-0.088

Figure 4 and 5, respectively present the normal probability plot of residuals for both linear and 2FI models of Ra. It can be seen all the residuals are falling on the straight line for both models. It indicates that errors are normally distributed, which satisfy the normality test conditions. Therefore, the model developed is adequate and valid. Figure 6 and 7 show a plot of experimental versus the predicted values of Ra. Since all the predicted values are close to the experimental values, confirming that the models could predict the responses accurately.

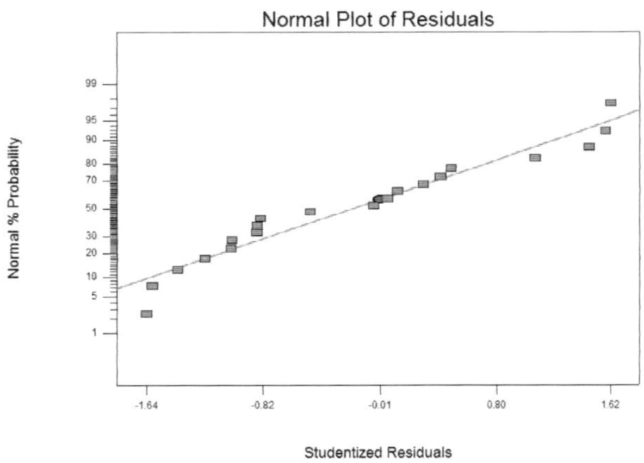

Figure 4. Normal probability plot for linear model

IOP Conf. Series: Journal of Physics: Conf. Series **892** (2017) 012002 doi:10.1088/1742-6596/892/1/012002

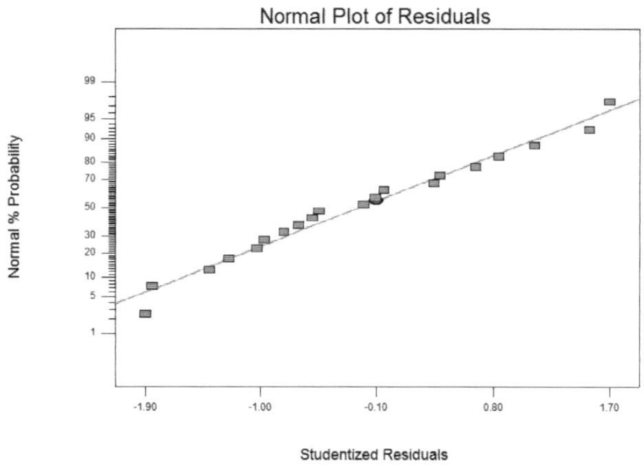

Figure 5. Normal probability plot for 2FI model

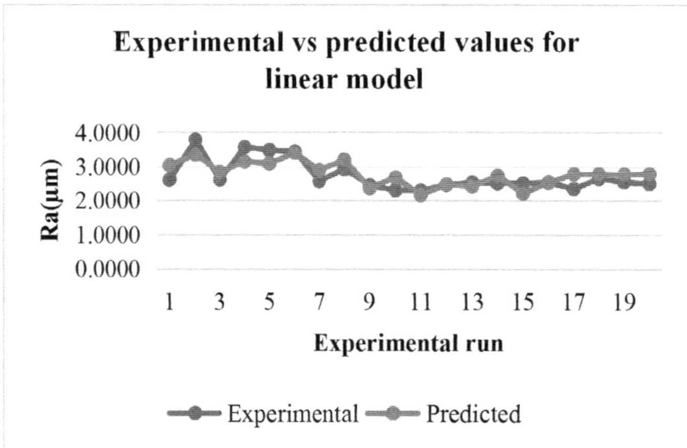

Figurecx 6. Experimental vs. Predicted values of the linear model

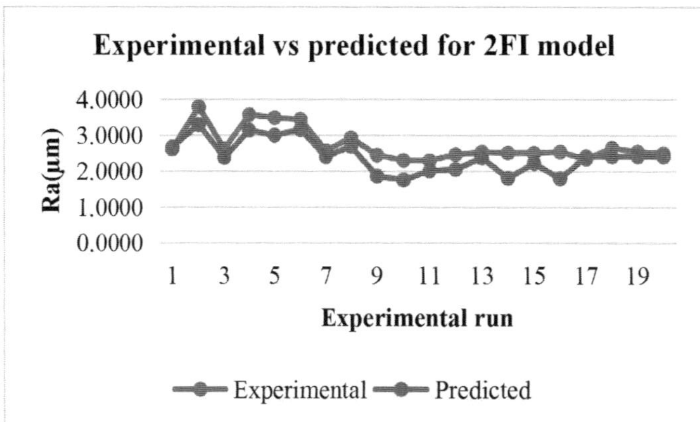

Figure 7. Experimental vs. Predicted values of the 2FI model

3.3 Confirmation test

Confirmation test was purposely performed to verify the adequacy and validation of the models developed within the given range of the machining parameters as presented in Table 2. Three sets of experiments were selected randomly. In order to estimate the accuracy of the prediction models, percentage error and average percentage error criteria were used as in (4):

$$PE(\%) = \frac{|\Pr edictedvalue - Experimentalvalue|}{Experimentalvalue} x100 \tag{4}$$

Where PE is the prediction error in %.

The predicted values and the actual experimental values were compared and their percentage error was calculated. The results of the confirmation test for Ra values are presented in Table 10. Prediction errors of linear model validations are found to be 16.57%, 16.31% and 11.49%. On the other hand, prediction errors of 2FI model are found to be 2.36%, 23.65% and 3.43%. Thus, the average prediction errors for both models are considered as 14.79% for the linear model and 9.81% for the 2FI model. The average percentage error on 2FI model of the Ra seems to be on slight lower rather than the linear model. And even some of the percentage error is in negative value.

Table 10. Results of the confirmation test.

Model	Machining parameters				Experimental value (μm)	Predicted value (μm)	% Error
	ON (μs)	OFF (μs)	IP (A)	SV (V)			
Linear	150	60	10	30	2.6106	3.0432	16.57
	230	60	10	60	2.3038	2.6796	16.31
	190	75	11	45	2.4989	2.7860	11.49
	Average percentage error (%)						14.79
2FI	150	60	10	30	2.6106	2.6722	2.36
	230	60	10	60	2.3038	1.7590	23.65
	190	75	11	45	2.4989	2.4131	3.43
	Average percentage error (%)						9.81

4. Optimization

Glowworm swarm optimization (GSO) algorithm is a nature inspired heuristics for optimization problem in swarm intelligence algorithms. Swarm intelligence is a type of artificial intelligence based on the collective behavior of decentralized, self-organized systems [25]. GSO algorithm was proposed by Krishnanand and Ghose in 2005 which is a derivative-free, meta-heuristic algorithm and mimicking the glow behavior of glowworms. The idea of the GSO algorithm comes from the biological behavior of the glowworms that attract mates or prey. The concept was like the brighter the glow, the more is the attraction. Moreover, GSO algorithm shares some general features with a well-known ant colony optimization (ACO) and particle swarm optimization (PSO), but with several significant differences and modifications [26].

The computational flowchart of the basic GSO algorithm can be summarized as in Figure 8 [27]:

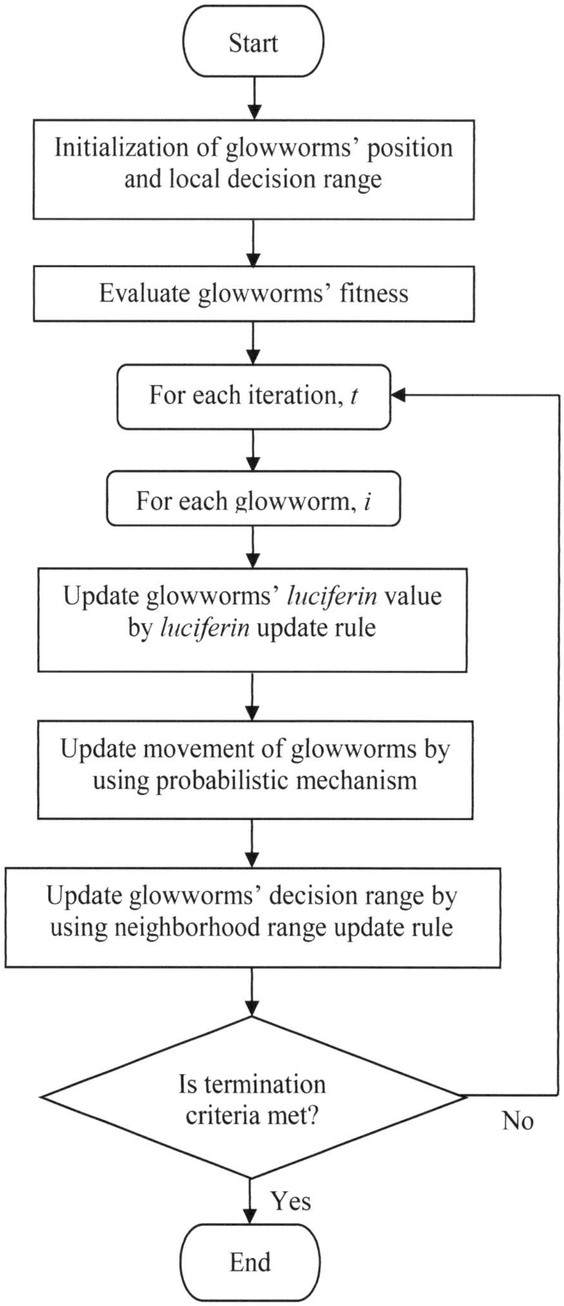

Figure 8. The flow chart of the GSO algorithm [27]

4.1 Optimization of surface roughness using GSO

The best selection of machining parameters improves not only on EDM cost, but also the surface integrity to a large extent by minimizing the roughness value of the work piece material. To evaluate the optimum performance of the GSO algorithm, the experimental data and mathematical modelling of two EDM processes are considered. Matlab R2014b was used for parametric optimization of the EDM processes with the following parameters: population size of glowworms *(n)* and the maximum iteration *(tmax)* are set to 20 and 100, respectively. Other parameter settings include initial luciferin value *(lo)* = 5, luciferin decay constant *(ρ)* = 0.4, luciferin enhancement constant *(γ)* = 0.6, beta (β) = 0.08, step size *(s)* = 0.03, neighbourhood range *($r_{id}(t)$)* = 3 and parameter used to control the number of neighbour *(nt)* = 5. In the present study, an effective GSO algorithm is developed to optimize the optimal combinations of machining parameters with the better surface roughness. The problem of optimization of machining parameters can be described by minimizing surface roughness as objective function. To formulate the objective function of optimization, both regression models which are given by equation (2) and (3), respectively were tested. Therefore, the objective function is written as follows:

$$minimize\ Ra\ (A,B,C,D) = 3.2756 + 0.0039A - 0.0068B + 0.0266C - 0.0225D \quad (5)$$

$$\begin{aligned} &minimize\ Ra\ (A,B,C,D) \\ &= -5.1249 + 0.0422A + 0.0174B + 0.8116C - 0.0431D + (5.8552E-05AB) \\ &- 0.0029AC = 0.0003AD - 0.0051BC + 0.0005BD + 0.0031CD \end{aligned} \quad (6)$$

Where Ra is surface roughness in µm, A is pulse on time in µs, B is pulse off time in µs, C is peak current in A and D is servo voltage in V.

The minimization of the objective function in (5) and (6) are subjected to the boundaries of the cutting condition values. Considering the lowest level is Level 1 and the highest level is Level 2, the limitation range of the optimization constraints are given as follows in (7)-(10):

$$150\mu s \leq A \leq 230\mu s \quad (7)$$

$$60\mu s \leq B \leq 90\mu s \quad (8)$$

$$10A \leq C \leq 12A \quad (9)$$

$$30V \leq D \leq 60V \quad (10)$$

The lowest and highest limitation range of cutting parameter is set as an input of GSO algorithm. The minimum Ra value is obtained at the last iteration *(tmax)* where the glowworms *(n)* that carry *luciferin (Ra)* being in the optimum position.

5. Results and discussion

Table 11 summarizes the results of the parametric optimization for consuming both linear and 2FI models of Ra as objective function with experimental results. The objective functions were optimized subject to feasible bound of the control variables as exhibited in (6)-(9). The table includes an optimum combination of machining parameters with the most minimal value of Ra. The minimum surface roughness value (2.2949µm) of the experiment was obtained at the combination of 150µs of pulse on time, 90µs of pulse off time, 10A of peak current and 60V of servo voltage. GSO optimization estimated the minimum surface roughness (2.2751µm) of linear model at the combination of 163.0319µs of pulse on time, 87.9741µs of pulse off time, 10.7982A of peak current and 58.9041V of servo voltage. While the 2FI model estimated the minimum value of Ra (2.0287µm) at the combination of 188.6636µs of pulse on time, 86.6291µs of pulse off time, 10.0605A of peak current and 59.3569V of servo voltage. Thus, it can be concluded that the 2FI model gives the most minimal value of Ra compared to the linear model and experimental result.

Table 11. Optimum cutting parameters and minimum surface roughness

Method	Optimum machining parameters				Minimum
	A (μs)	B (μs)	C (A)	D (V)	Ra (μm)
Experimental	150	90	10	60	2.2949
Linear model	163.0319	87.9741	10.7982	58.9041	2.2751
2FI model	188.6636	86.6291	10.0605	59.3569	2.0287

6. Conclusion

In this study, the effects of the pulse on time (ON), pulse off time OFF), peak current (IP) and servo voltage (SV) on the Ra of die sinking EDM was studied by using integrated regression and glowworm swarm optimization algorithm. The experimental data were collected from AG40L Sodick electrical discharge machine with full factorial design of experiment with added four center points. The validity of the linear and 2FI models of Ra was performed by using ANOVA and F test. Then, both models were applied as the objective function for the GSO optimization process. Statistical analysis was used as to discuss the results from the optimization process. From this integrated study, the following conclusions are made:

- Ra has found to be most minimum at the run no. 11 (2.2949μm) which the combinations of the cutting parameters are 150μs of pulse on time, 90μs of pulse off time, 10A of peak current and 60V of servo voltage.
- The predicted Ra values match the experimental values reasonably well, with R-squared of 0.608 and 0.835 for linear model and 2FI model, respectively.
- Based on the confirmation test, 2FI model of Ra gives slightly lower of the percentage error compared to the linear model of Ra.
- 2FI model of Ra predicted value of Ra (2.0287μm) much lower rather than the linear model (2.2751μm).
- For optimization process, it can be summarized that the 2FI model of Ra gives the most minimal value of Ra compared to the linear model and experimental result.

Acknowledgement

Special appreciation to reviewer(s) for useful advices and comments. The authors greatly acknowledge the Research Management Centre, UTM and Ministry of Higher Education Malaysia (MOHE) for financial support through the Tier 1 vot. No. QJ130000.2528.11H72/Tier1.

References

[1] Nandi A K and Pratihar D K An expert system based on FBFN using a GA to predict surface finish in ultra-precision turning 2014 *Journal of materials processing technology* **155** p 1150-56
[2] Deris A M, Zain A M and Sallehuddin R Hybrid GR-SVM for prediction of surface roughness in abrasive water jet machining 2013 *Meccanica* **48** p 1937-45
[3] A. M. Zain, H. Haron, and S. Sharif Integrated ANN–GA for estimating the minimum value for machining performance 2012 *Int. Journal of Production Research* **50** p 191-213
[4] Zainal N, Zain A M, Radzi N H M, and Udin A Glowworm swarm optimization (GSO) algorithm for optimization problems: A state-of-the-art review 2013 *In Applied Mechanics*

and Materials, Trans Tech Publications **421** p 507-511
[5] Samah A A, Zainudin Z, Majid H A, and Yusoff S N M A Framework using an Evolutionary Algorithm for On-call Doctor Scheduling 2012 *Journal of Computer Science & Computational Mathematics* **2** p 9-16
[6] Zain A M, Haron H, and Sharif S Review of ANN technique for modeling surface roughness performance measure in machining process 2009 *3rd Asia Int. Conf. on Modelling & Simulation* p 35-39
[7] Chandrasekaran V, Kanagarajan D and Karthikeyan R Optimization of EDM Characteristics of WC/5ni Composites Using Response Surface Methodology 2013 *Int. Journal of Recent Technology and Engineering (IJRTE) ISSN* **2** p 2277-3878
[8] Deris A M, Zain A M and Sallehuddin R Overview of support vector machine in modeling machining performances 2011 *Procedia Engineering* **24** p 308-312
[9] Zain A M, Haron H and Sharif S Integration of simulated annealing and genetic algorithm to estimate optimal solutions for minimising surface roughness in end milling Ti-6AL-4V 2011 *Int. Journal of Computer Integrated Manufacturing* **24** p 574-592
[10] Zainal N, Zain A M and Sharif S A study of electrode wear ratio on EDM of Ti-6AL-4V with copper-tungsten electrode 2016 *MATEC Web of Conferences, EDP Sciences* **78**
[11] Wang K, Gelgele H L, Wang Y, Yuan Q and Fang M A hybrid intelligent method for modelling the EDM process 2003 *Int. Journal of Machine Tools and Manufacture* **43** p 995-999
[12] Padhee S, Nayak N, Panda S K, Dhal P R and Mahapatra S S Multi-objective parametric optimization of powder mixed electro-discharge machining using response surface methodology and non-dominated sorting genetic algorithm 2012 *Sadhana* **37** p 223-240
[13] Singh G and Dhiman D P Review: Parametric Optimization of Edm Machine using Taghuchi & Anova Technique 2016 **3** p 783-788
[14] Li C P, Kim M Y, Islam M M and Ko T J Mechanism analysis of hybrid machining process comprising EDM and end milling 2016 *Journal of Materials Processing Technology* p 309-319
[15] Baraskar S S, Banwait S S and Laroiya S C Mathematical Modeling of Electrical Discharge Machining Process through Response Surface Methodology 2011 **2** p 1-10
[16] Baraskar S S, Banwait S S and Laroiya S C Multiobjective Optimization of Electrical Discharge Machining Process Using a Hybrid Method 2012 *Journal of Materials and Manufacturing Process* **28** p 348-354
[17] Razak N H, Rahman M M, Noor M M and Kadirgama K Artificial Intelligence Techniques For Machining Performance: A Review 2010 p 39-53
[18] Mohamad A, Zain A M, Bazin N E N and Udin A A process prediction model based on Cuckoo algorithm for abrasive waterjet machining 2015 *Journal of Intelligent Manufacturing* **26** p 1247-52
[19] Patel K M, Pandey P M and Rao P V Determination of an optimum parametric combination using a surface roughness prediction model for EDM of Al2O3/SiCw/TiC ceramic composite 2009 *Materials and Manufacturing Processes* **24** p 675-682
[20] Bhattacharyya B, Gangopadhyay S and Sarkar B R Modelling and analysis of EDM ED job surface integrity 2007 *Journal of Materials Processing Technology* **189** p 169-177
[21] Mandal A, Dixit A R, Das A K and Mandal N Modeling and Optimization of Machining Nimonic C-263 Superalloy using Multicut Strategy in WEDM 2015 *Journal of Materials and Manufacturing Processes* **31** p 860-868
[22] Gopalakannan S, Senthilvelan T and Ranganathan S Modeling and Optimization of EDM Process Parameters on Machining of Al 7075-B 4 C MMC Using RSM 2012 *Procedia Engineering* **38** p 685-690
[23] Anitha J, Das R and Pradhan M K Multi-Objective Optimization of Electrical Discharge Machining Processes Using Artificial Neural Network 2016 *Journal of Mechanical and Industrial Engineering* **10** p 11-18

[24] Zain A M, Haron H, Qasem S N and Sharif S Regression and ANN models for estimating minimum value of machining performance 2012 *Applied Mathematical Modelling* **36** p 1477-92
[25] Liao W, Kao Y and Li Y A sensor deployment approach using glowworm swarm optimization algorithm in wireless sensor networks 2011 *Journal of expert system with applications* **38** p 12180-88
[26] Wu B, Qian C, Ni W and Fan S The improvement of glowworm swarm optimization for continuous optimization problems 2012 *Journal of Expert System with Applications* **39** p 6335-42
[27] Zainal N, Zain A M, Radzi N H M and Othman M R Glowworm swarm optimization (GSO) for optimization of machining parameters 2014 *Journal of Intelligent Manufacturing* **27** p 797-804

IOP Conf. Series: Journal of Physics: Conf. Series **892** (2017) 012003 doi:10.1088/1742-6596/892/1/012003

Harmony search optimization in dimensional accuracy of die sinking EDM process using SS316L stainless steel

A M Deris[1], A M Zain[1], R Sallehuddin[1], S Sharif[2]

[1]Soft Computing Research Group, Faculty of Computing, UTM Johor Bahru, Malaysia
[2]Faculty of Mechanical Engineering, UTM Johor Bahru, Malaysia

azlanmz@utm.my

Abstract. Electric discharge machine (EDM) is one of the widely used nonconventional machining processes for hard and difficult to machine materials. Due to the large number of machining parameters in EDM and its complicated structural, the selection of the optimal solution of machining parameters for obtaining minimum machining performance is remain as a challenging task to the researchers. This paper proposed experimental investigation and optimization of machining parameters for EDM process on stainless steel 316L work piece using Harmony Search (HS) algorithm. The mathematical model was developed based on regression approach with four input parameters which are pulse on time, peak current, servo voltage and servo speed to the output response which is dimensional accuracy (DA). The optimal result of HS approach was compared with regression analysis and it was found HS gave better result y giving the most minimum DA value compared with regression approach.

1. Introduction

Manufacturing can be defines as the integration of the complex process of mechanical, physical and chemical to make finish part of products [1]. In manufacturing process, machining is the most important and widely used in manufacturing process instead of molding, casting and joining as the machining process can gives the closer dimensional and also can performed the external and internal geometric features as well as sharp corners and flatness. Machining generally can be defined as a process of material removing from the work piece in a form of chips. Machining consists of three major which are machine tools, cutting tool and work piece. There are two types of machining; conventional and modern machining. Conventional machining consists of traditional process work piece removal in the form of chips such as turning, milling, grinding and boring while modern machining comes in terms of chemical items or advanced technologies such as Electric Discharge Machining (EDM), wire electric discharge machining (WEDM), electrochemical machining (ECM), abrasive water jet (AWJ) and laser beam machining (LBM) [2]-[4]. EDM is one of the mostly used modern machining processes for hard material and complex shapes which are difficult-to-machine by conventional machining. EDM technology is a reliable, affordable and accurate process which is commonly used in automobile, surgical industries, molds, and aerospace fields. EDM has unique feature that differs from other machining process. The direct contact does not occurred in EDM during the cutting process between the work piece and electrode when eliminating mechanical stresses, chatter and vibration problems. EDM technology has been improve significantly and has been developed in many ideas especially in the manufacturing industries that yielded enormous benefits in economic as well as generating keen interest in research area. There are different types of EDM that have been interest by researchers including die-sinking EDM, wire EDM (WEDM), powder-mixed EDM, Dry EDM and Micro-EDM [5]. For die sinking EDM, the machining process occurs by the 3D movement of an electrode with the numerical control monitors the gap conditions and synchronously controls the different axes and the pulse generator. In WEDM, the material is eroded from the work piece by a series of discrete sparks occurring between the work piece and the wire separated by the existence of dielectric fluid. Micro-EDM is a process that is capable not only for micro-holes and micro-shafts with 5 μm in diameter but also capable for complex 3D micro-cavities. Lastly, powder-mixed EDM is basically differs from conventional EDM when a suitable material in powder form is mixed into dielectric fluid. The spark gap filled up additive particles when a suitable voltage is applied and the gap distance setup between tool and work piece increase from 25-50 to 50-150 μm.

There are several machining performances that are evaluated by major measure, such as cutting force, tool wear, material removal rate, tool life, chip form and surface finish [6]. Dimensional accuracy is also one of the important parameters that are considered by many researchers [7] – [8]. Dimensional Accuracy (DA) which also can be defines as diameter overcut is significantly influences the precision and accuracy of the product. Dimensional accuracy is crucial when close tolerance components are required to be produced for space applications and tools, plastic molding and die casting. Tiwari et al. [7] optimized four machining parameters which are electrolytic concentration, voltage, feed rate and inter electrode gap using Taguchi method in order to find the optimal solution of overcut in ECM process. The experiment was conducted based on design of experiment L27 orthogonal array using EN19 tool steel material. The mathematical model was developed using regression analysis and it was found that voltage and concentration were the most influence machining parameters for overcut value. Pradhan and Das [8] investigate radial overcut in EDM process based on general regression neural network using AISI D2 tool steel. A full factorial design was used to conduct the experiments. The result shows that the proposed models can be successfully employed to predict the EDM process. Thanigaivelan and Arunachalam [9] investigated the effect and parametric optimization of machining rate and overcut of electrochemical micromachining based on grey relational analysis. the experiments is conducted using stainless steel 304 material with four machining parameters, which are tool tip shape, voltage, pulse on time and electrolyte concentration. The result found that the optimization of the multiple performances can be simplified and improved through Taguchi and grey relational analysis approach.

A new trend of optimization process using artificial intelligent approach has been evolves significantly over the years [2]. Meta-heuristic approaches such as artificial bee colony (ABC), simulated annealing (SA), ant colony optimization (ACO), bat algorithm (BA), firefly algorithm (FA), cuckoo search algorithm (CS) and harmony search (HS) optimization among the interest of many researchers [10]-[14]. These optimization approaches are proven to give better results compared with conventional optimization approaches. Yusup et al [15] investigate the optimal machining parameters of AWJ machining using artificial bee colony and it was found that ABC approach gave better results compared with other method such as regression and artificial neural network (ANN). Rao and Venkaiah [16] optimized WEDM parameters of niminic-263 alloy using particle swarm optimization. The mathematical model for material removal rate (MRR) and surface roughness (Ra) were developed based on response surface methodology (RSM). The result shows that PSO gave better performance compared with RSM. Raja et al. [17] optimized the EDM parameters on hardened die steel using firefly algorithm and it was found that FA is suitable for solving machining parameters optimization problem as the proposed model reduces time and cost of machining trials for surface roughness prediction. Based on the review, the optimization of EDM parameters based on Harmony Search approach is still lacking, hence this paper proposed HS approach in order to find the optimal solution of dimensional accuracy parameters in EDM process. The mathematical model was developed based on multi linear regression approach [18, 19]. The development of mathematical model for DA is important in order to have better understanding to the machining process [20].

IOP Conf. Series: Journal of Physics: Conf. Series **892** (2017) 012003 doi:10.1088/1742-6596/892/1/012003

2. Experimental Procedure

The experiment is conducted using AG40L die sinking EDM as shown in Figure 1. The material used in this experiment is stainless steel 316L as shown in Figure 2, and the electrode is copper impregnated graphite is shown in Figure 3. The details chemical composition and mechanical properties of SS 316L are shown in Table 1 and Table 2 respectively.

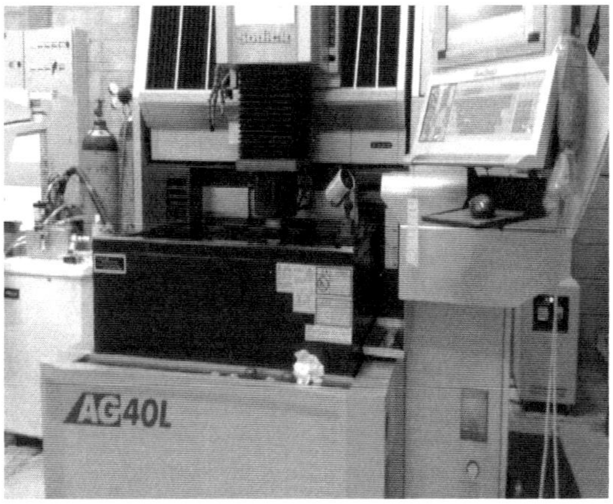

Figure 1. AG40L model of die sinking EDM

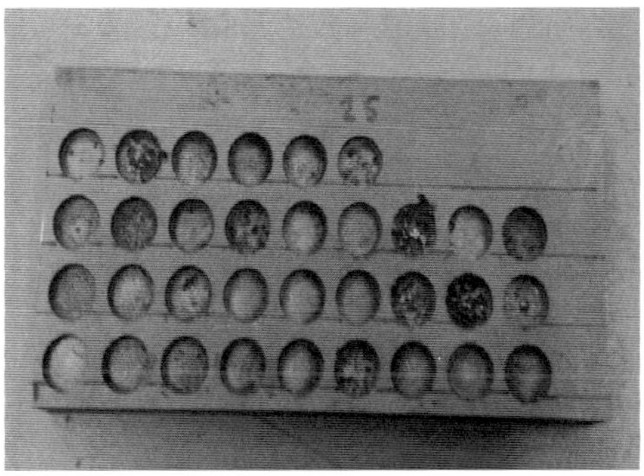

Figure 2. Stainless steel 316L

Figure 3. Copper impregnated carbide electrode

IOP Conf. Series: Journal of Physics: Conf. Series **892** (2017) 012003 doi:10.1088/1742-6596/892/1/012003

Table 1. Chemical composition of SS 316L

Elements	316L (wt %)
C	0.026
Si	0.37
Mn	0.16
Cr	16.55
Cu	0.16
Ni	10.0
P	0.029
S	0.027
Mo	2.02
N	0.036
Fe	Balance

Table 2. Mechanical properties of SS 316L

Mechanical Properties	Typical	Minimum
Tensile Strength	600MPa	485MPa
Proof Strength, (offset 0.2%)	310MPa	170MPa
Elongation (Percent in 50mm)	60	40
Hardness (Brinell)	217	-
Hardness (Rockwell)	95	-
Endurance (Fatigue Limit)	240MPa	-

The design setting is important process that needs to be completed prior to the experimental process can be run. In this study, the experiment is conducted based on two levels full factorial design of four input variables. Design of Experiment (DOE) is implemented to observe the process characteristic since it provides the best parameters of EDM to fulfill the machining objectives. DOE is a systematic and powerful tool that applies principles and techniques at the data collection stage, to ensure the generations of valid, defensible and supportable engineering conclusions. DOE can observe and identify the change that has been made from input variables. The experiment implies 2 levels of 4 input factors which consist of 16 runs (2^4). Before conducting the experiment, the range for low and high level for each factor is determined. In this study the selection of the range for EDM parameters is based on manual EDM handbook and recommendation from previous researchers. Table 3 shows the range value of EDM parameters.

Table 3. The range value of EDM parameters

Variables	Unit	Levels	
		1	2
Pulse on time (ON)	μs	100	200
Peak Current (IP)	A	5.7	10.5
Servo Voltage (SV)	V	30	90
Servo Speed (S)	-	74	92

IOP Conf. Series: Journal of Physics: Conf. Series **892** (2017) 012003 doi:10.1088/1742-6596/892/1/012003

Dimensional Accuracy (DA) which also can be defines as diameter overcut is one of the important responses in machining process. Dimensional accuracy is crucial when close tolerance components are required to be produced for space applications and tools, plastic molding and die casting. Dimensional accuracy can be calculated as follows:

$$DA = (D_b - D_a) / D_b \ (\%) \qquad (1)$$

Where D_b is the diameter electrode before the machining and D_a is the diameter electrode after the machining.

3. Experimental Result and Analysis

The machining experimental has been conducted in in full factorial design which given 16 number of runs. Figure 4 shows the materials cutting after the machining process while Table 4 shows the experimental result of DA value.

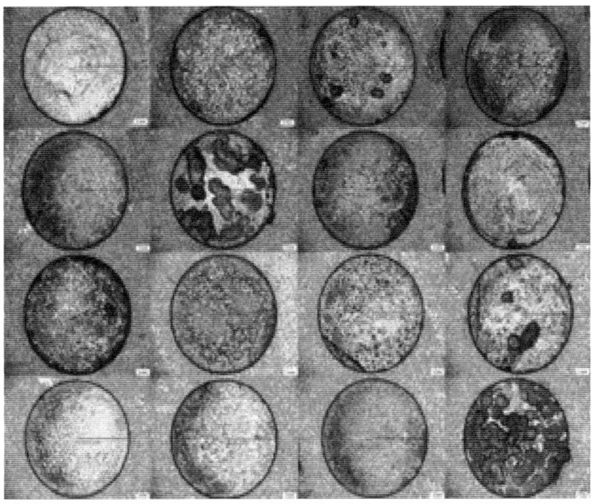

Figure 4. Holes of materials after machining

Table 4: Experimental data

No.	ON	IP	SV	S	DA
1	100.00	5.70	30.00	74.00	0.023
2	100.00	10.50	30.00	74.00	0.0223
3	100.00	5.70	90.00	74.00	0.0158
4	100.00	10.50	90.00	74.00	0.0280
5	100.00	5.70	30.00	92.00	0.0133
6	100.00	10.50	30.00	92.00	0.0275
7	100.00	5.70	90.00	92.00	0.0150
8	100.00	10.50	90.00	92.00	0.0228
9	200.00	5.70	30.00	74.00	0.0223
10	200.00	10.50	30.00	74.00	0.0240
11	200.00	5.70	90.00	74.00	0.0125
12	200.00	10.50	90.00	74.00	0.0238
13	200.00	5.70	30.00	92.00	0.0238
14	200.00	10.50	30.00	92.00	0.0303
15	200.00	5.70	90.00	92.00	0.0183

16	200.00	10.50	90.00	92.00	0.0243

Based on the table, it can be found that the most minimum Da value is 0.0125. The regression model was developed based on experimental data and analyzed using analysis of variance (ANOVA). Sum square (SS), mean square (MS), F-ratio and P-values associated with each other levels and interaction were presented in Table 5. From the table, it shown that the model is statistically significant with the P-value Prob >F is 0.0157, which is less than 0.05, while peak current is the only parameter that is significant to the dimensional accuracy value with the p-value of 0.0021. The rest of the parameters show insignificant with the p-value are greater than 0.05. Figure 5 shows the graphical plots if normal plot of residuals for dimensional accuracy in EDM process.

Table 5. ANOVA analysis of DA value

Source	SS	DF	MS	F	P-value
Model	2.690E-004	4	6.726E-005	4.96	0.0157
ON	8.410E-006	1	8.410E-006	0.62	0.4477
IP	2.176E-004	1	2.176E-004	16.04	0.0021
SV	4.225E-005	1	4.225E-005	3.11	0.1053
S	8.100E-007	1	8.100E-007	0.060	0.8114
Residual	1.492E-004	11	1.356E-005		
Cor Total	4.182E-004	15			

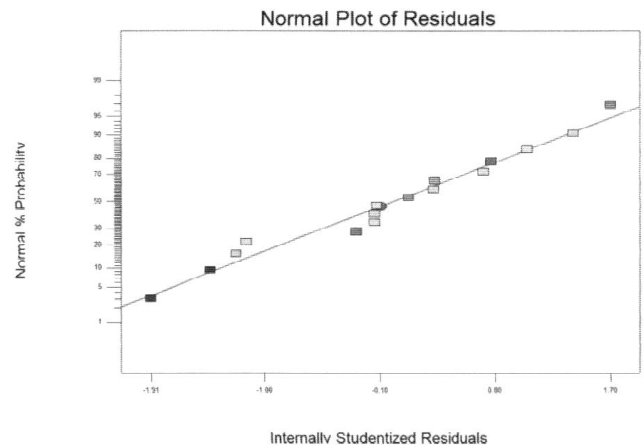

Figure 5 . Normal plot of residual for DA

Based on the plots, it can be seen that all the points are scattered normally and distributed along the line as well as no abnormality found. There are no obvious patterns even though some data points are far from the centre line and all the results are within the acceptable range. These are the desirable plots when analyzing the data. Therefore, it can be concluded that the model was valid and adequate. The multiple linear regression for DA can be stated as in Equation 2:

$$DA = \beta_0 + \beta_1 x_1 + \beta_2 x_2 + \beta_3 x_3 + \beta_4 x_4 + \beta_5 x_1 x_2 + \beta_6 x_1 x_3 + \beta_7 x_1 x_4 + \beta_8 x_2 x_3 + \beta_9 x_2 x_4 + \beta_{10} x_3 x_4 \quad (2)$$

Where $\boldsymbol{\beta_n}$ are the regression coefficients and $\boldsymbol{x_n}$ are the variable of machining parameters. Based on the experimental data, the final mathematical model for DA can be stated as in Equation 3:

DA = +8.24219E-003 +1.45000E-005 * ON +1.53646E-003 * IP -5.41667E-005 * SV +2.50000E-005*S (3)

4. Harmony Search Optimization

Harmony Search (HS) [21] is a new meta-heuristic algorithm mimicking the improvisation of music players in searching for a perfect state of harmony. The optimization process is applied when musician plays different notes of different musical instrument to find the best combination of frequency for a best tune [22]. The aesthetic quality of music produced by its musical instruments can be determined by its pitch, timbre and amplitude. HS has simple mathematical model and only few lines of algorithm involve. The objectives function for optimization problems are normally stated as follows:

$$\text{Minimize } f(x), \quad \text{Subject to } x_i = X_i, i = 1,2,3, \dots n \tag{4}$$

where *f(x)* is the objective function; *x* is the set of each decision variable x_i; *n* is the number of decision variables, *Xi* is the set of the possible range of values for each decision variable, that is $x_i^L \leq X_i \leq x_i^U$ and x_i^L and x_i^U are the lower and upper bound for each decision variable. Steps for HS algorithm describes as follow:

Step 1: Initialize the parameters.

i. The HS algorithm parameters are specified in this step. These parameters are:
ii. Harmony Memory Size (HMS), or the number of solution vectors in the harmony memory
iii. Harmony Memory Considering Rate (HMCR), HMCR $\in [0,1]$ to determine the rate of selecting the value from the memory;
iv. Pitch Adjusting Rate (PAR) $\in [0,1]$ which determines the probability of local improvement
v. The fret width (FW) which determines the distance of adjustment
vi. Number of improvisations (NI) or number of iterations.

Step 2: Initialize the Harmony Memory (HM)

HM is a repository of the population individuals, HM = $[x^1, x^2 \dots x^{HMS}]^T$ of the size HMS. In this step, these individuals are randomly generated as $x_i^j = LB_i + (UB_i - LB_i) \text{ x } U(0,1), \forall = 1,2, \dots, N$ and $\forall j = 1,2,..HMS,$ and U (0, 1) generates a uniform random number between 0 and 1.

Step 3: Improvise a new harmony.

A New Harmony vector is generated, $x' = (x_i', x_2', \dots, x_N')$, based on three operators: (i) memory consideration (MC), (ii) pitch adjustment (PA), and (iii) random consideration (RC). The three operators assign a value for each decision variable x_i' in the New Harmony as formulated in Equation 5 as follows:

$$x_i^l \leftarrow \begin{cases} x_i' \in \{x_i^1, x_i^2, \dots x_i^{HMS} & \text{w.p HMCR x } (1 - PAR)\ \{MC\} \\ x_i' = x_i' + U(-1,1)x\ FW & \text{w.p. HMCR x } PAR\ \ \{PA\} \\ x_i' \in X_i & \text{w.p } (1 - HMCR)\ \ \{RC\} \end{cases} \tag{5}$$

Step 4: Update the HM

The new harmony vector,$x' = (x_i', x_2', \ldots, x_N')$, replaces the worst harmony stored in HM.

Step 5: Check the stop criterion

Step 3 and step 4 of HS algorithm are repeated until the stop criterion (which is normally depends on NI) is met.

The optimization process of harmony search is conducted using Matlab software in order to find the optimal solution of dimensional accuracy. HS is expected to give the minimum value of DA for EDM process. Figure 6 shows the flowchart of HS optimization while the result of HS optimization is stated in Table 6.

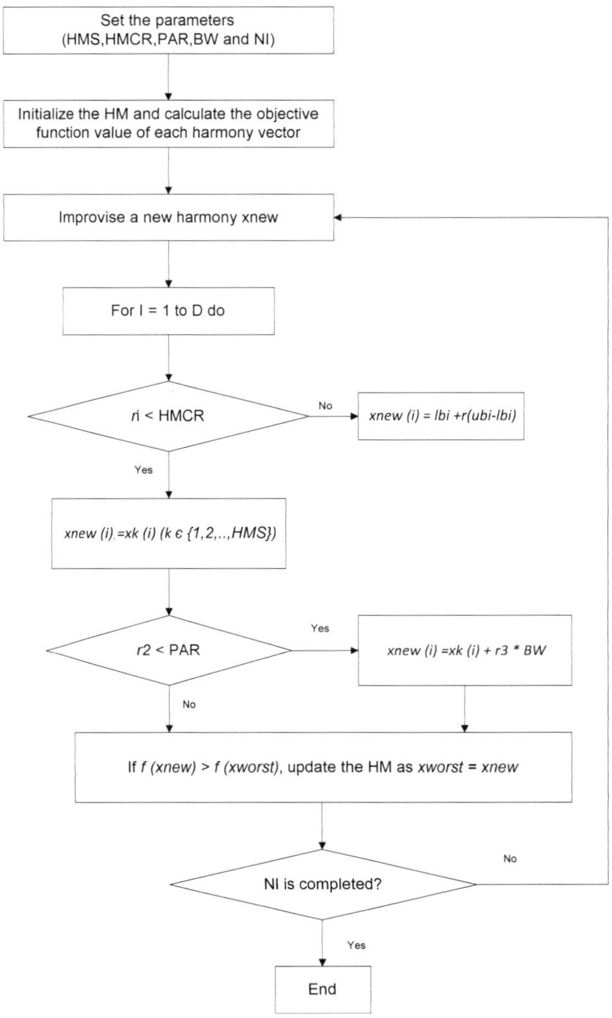

Figure 6. Flowchart of HS Optimization

Table 6. Optimization results for dimensional accuracy

	Optimal cutting points (ON,OFF,IP,SV,S)	Minimum DA (%)	Percentage Error (%)
Experimental	100,50,5.7,30,74	0.0125	-

MLR	100,50,5.7,30,74	0.0168	0.34 %
HS	110.0211 5.7000 89.1258 74.0169	0.0156	0.25 %

Table 6 shows the results of multi linear regression and HS optimization compared with experimental data. The percentage error of multi linear regression and HS were calculated in order to find the best approach for DA optimization of EDM process. Based on the result, it can be seen that Harmony Search optimization outperformed multi linear regression by 0.25% of percentage error compared with 0.34% of percentage error. The optimal solution can be achieved when pulse on time 110.0211µs, peak current 5.7A, servo voltage 89.1258V and servo speed is 74.0169, which giving the most minimum DA value is 0.0156%.

5. Conclusion

Machining experiment was conducted using stainless steel 316L and copper impregnated carbide electrode for optimization of dimensional accuracy parameters in electric discharge machining process. The experiment was conducted based on full factorial design with four machining parameters which are pulse of time, peak current, servo voltage and servo speed. The mathematical model was developed based on multi liner regression approach and the model was analyzed using ANOVA analysis. ANOVA and regression approach provide a systematic and effective methodology for the optimization process. It is found that peak current is the most influence parameters to the dimensional accuracy value compared with pulse on time, servo voltage and servo speed.

The optimization process is conducted using harmony search approach in order to find the optimal solution of dimensional accuracy for EDM process. The minimum DA value of harmony search then compared with multi linear regression and it was found that HS gave the better results by giving the less percentage error. Hence, optimization process using HS approach is the most effective method compared with conventional optimization techniques such as ANOVA and regression approaches.

Acknowledgement

Special appreciative to reviewer(s) for the useful advices and comments. The authors greatly acknowledge the Research Management Centre, UTM for financial support through the *Research University Grant* (*GUP) Vot. No.* Q.J130000.2528.14H14.

References

[1] Rao R V 2010 Advanced modeling and optimization of manufacturing processes: international research and development. *Springer series in adv manuf.* 2010.

[2] Zain A M, Haron H and Sharif S 2011 Estimation of the minimum machining performance in the abrasive water jet machining using integrated ANN-SA. *Exp Sys Appl* 38, 8316-8326

[3] Tang L and Gan W M 2014 Utilization of flow field simulations for cathode design in electrochemical machining of aerospace engine blisk channels *Int J Adv Manuf Techno* 72, 1759-1766

[4] Ay M, Çaydaş U and Hasçalık A 2013 Optimization of micro-EDM drilling of Inconel 718 superalloy. *Int J Adv Manuf Techno* 66, 1015-1023

[5] Garg R K, Singh K K, Sachdeva A, Sharma V S, Ojha K and Singh S 2010 Review of research work in sinking EDM and WEDM on metal matrix composite materials *Int J Adv Manuf Techno* 50, 611-624

[6] Jawahir I S, Balaji A K, Rouch K E and Baker J R 2003 Towards integration of hybrid models for optimized machining performance in intelligent manufacturing systems. *J Mater Process Techno* 139, 488–498
[7] Tiwari A, Mandal A and Kumar K 2015 Optimization of Overcut in Electrochemical Machining for EN 19 Tool Steel Using Taguchi Approach. *Mater Today: Proceedings* 2, 2337-2345
[8] Pradhan M K and Das R 2015 Application of a general regression neural network for predicting radial overcut in electrical discharge machining of AISI D2 tool steel." *Int J Machining Machinability Mater* 17, 355-369
[9] Thanigaivelan R and Arunachalam R 2013 Optimization of process parameters on machining rate and overcut in electrochemical micromachining using grey relational analysis *J Science Industrial Reseach* 72, 36-42
[10] Zain A M, Haron H and Sharif 2011 Genetic algorithm and simulated annealing to estimate optimal process parameters of the abrasive waterjet machining *Eng Comp* 27, 251-259
[11] Yusup N, Zain A M and Hashim S Z M 2012 Overview of PSO for optimizing process parameters of machining *Procedia Eng* 29, 914-923.
[12] Mohamad A B, Zain A M and Bazin, N E B 2014 Cuckoo search algorithm for optimization problems—a literature review and its applications *Appl Artif Intell* 28, 419-448
[13] Teimouri R and Baseri H 2014 Optimization of magnetic field assisted EDM using the continuous ACO algorithm *Appl Soft Comp* 14, 381-389
[14] Majumder A 2013 Process parameter optimization during EDM of AISI 316 LN stainless steel by using fuzzy based multi-objective PSO *J Mech Sci Techno* 27, 2143-2151
[15] Yusup N, Sarkheyli A, Zain A M, Hashim S Z M and Ithnin N 2014 Estimation of optimal machining control parameters using artificial bee colony *J Intell Manuf* 25, 1463-1472
[16] Rao M S and Venkaiah N 2015 Parametric Optimization in Machining of Nimonic-263 Alloy using RSM and Particle Swarm Optimization *Int Conf Nanomater Techno* 10, 70-79
[17] Raja S B, Pramod C V S, Krishna K V, Ragunathan A and Vinesh S 2015 Parametric Optimization of electrical discharge machining parameters on hardened die steel using firefly algorithm *Eng Comp* 31, 1-9
[18] Olive, D.J., 2017. Multiple Linear Regression in Linear Regression. Springer International Publishing, 17-83.
[19] Łyko J and Rudek R 2016 *An Algorithmic Analysis of Degressively Proportional Apportionments. J Comp Sci & Comp Math 6, 53-57.*
[20] Zain A M, Haron H and Sharif S 2010 Application of GA to optimize cutting conditions for minimizing surface roughness in end milling machining process. *Exp Sys Appl* 37, 4650-4659
[21] Geem Z W, Kim J H and Loganathan G V 2001 A new heuristic optimization algorithm: harmony search *Simulation* 76 60-68
[22] Patil S A and Patel D A 2013 An overview: Improved harmony search algorithm and its applications in mechanical engineering *Int J Eng Sci Innov Techno* 2 433-444

IOP Conf. Series: Journal of Physics: Conf. Series **892** (2017) 012004 doi:10.1088/1742-6596/892/1/012004

A Review On Missing Value Estimation Using Imputation Algorithm

Roslan Armina[1], Azlan Mohd Zain[2] ,Nor Azizah Ali[3] and Roselina Sallehuddin[4]

Soft Computing Research Group, Faculty of Computing, Universiti Teknologi Malaysia, Malaysia
E-mail: roslanarmina@gmail.com, azlanmz@utm.my, nzah@utm.my, roselina@utm.my

Abstract. The presence of the missing value in the data set has always been a major problem for precise prediction. The method for imputing missing value needs to minimize the effect of incomplete data sets for the prediction model. Many algorithms have been proposed for countermeasure of missing value problem. In this review, we provide a comprehensive analysis of existing imputation algorithm, focusing on the technique used and the implementation of global or local information of data sets for missing value estimation. In addition validation method for imputation result and way to measure the performance of imputation algorithm also described. The objective of this review is to highlight possible improvement on existing method and it is hoped that this review gives reader better understanding of imputation method trend.

1. Introduction

Data quality is considered important and concerned as big role in Machine learning and other correlated area Knowledge Discovery from databases. In Machine Learning, most of algorithm run the data mining straight from the data sets to extract knowledge and the quality of knowledge are depend on the quality of the data[1]
The presence of missing value is the common problem for all type research. Unfortunately, to acquire the actual data of missing value is often impossible and hence the data analyst must accept with reality that some data is unavailable. The missing value presence in unavoidable problem in many term because most of the data mining algorithm could not work with data sets that consist missing value[2]

2. The Imputation Technique Of Missing Value

Imputation method involve the replacement of missing value with some value that has been estimated that based on data mining of available information in the data set[2]. There are several methods available for handling the missing value which are Case Deletion (CD) method that consist the missing value determination on each of attribute and instance, Parameter Estimation method that use the application of procedures of maximum likelihood procedures for estimation of the parameters of a model defined for the complete data. Expectation-Maximization algorithm is applied in maximum likelihood procedures [3], and imputation method that constitute of class of procedures that solely aims for replacement of missing values with estimated values. The employment of known relationships that can be identified in the valid values of the data set to assist in estimating the missing values is the objective of this method. This papers focus on imputation of missing data.

2.1 Global Approach

In this global approach, Missing value replacement algorithm is based on the algorithm that apply on global correlation information that originated from derivation of all data matrix. The existence of a global covariance structure has been assume among all genes or samples in the expression matrix[4]. When this assumption is not appropriate, as example their imputation becomes less precise when the genes exhibit dominant local similarity structures. The widely used algorithm for this imputation method category is Bayesian principal component analysis (BPCA) [5] and Singular Vector Decomposition imputation (SVDimpute) [6]. For the SVDimpute algorithm, the Singular Vector Decomposition is used to obtain eigengenes that is a set of pattern of mutually orthogonal expression, which can be linearly combined to approximate the expression of all genes in the data set. The Singular Vector Decomposition algorithm can only be run on complete matrices[7].First, SVDimpute algortihm regresses the gene against the most significant eigengenes k most significant and to reconstruct the missing values from a linear combination of the k eigengenes ,coefficients of the regression is applied.

2.2 Local Approach

In this local approach, for missing value imputation application, the algorithms exploit only local similarity structure in the data sets. To computes the missing value genes that contain the missing values are used with the gene subsets that show high correlation. The example of local approach such as the K-Nearest Neighbour imputation (KNNimpute) and local least square imputation (LLSimpute) are some of the earliest and well-known algorithms [8]. The k-nearest neighbor imputation (KNNimpute) [9] uses pairwise information between the target gene with missing values and the k nearest reference genes to impute the missing values. The research before have shown that KNNimpute imputation algorithm performs extremely well when strong correlation exists between genes in the data sets .For imputes missing values,Local Least Square imputation LLSimpute [10] apply a multiple regression model. This technique has been proven to be slightly more competitive than KNNimpute and is more complex than BPCA.

SLLSimpute sequentially performs imputation by starting from the least missing gene with least missing rates.Sequential LLSimpute (SLLSimpute) [11] is an extension of LLSimpute algorithm and the gene that has been imputed are then reused for imputation of other genes. It has been proven that SLLSimpute performs better than LLSimpute because the genes with missing values are reusable in this algorithm. Another method that widely used in local approach is the MICE-CART, which is the combination of multiple imputation of classification and regression trees (CART) and chained equations (MICE). It is a nonparametric approach done by [12] to perform multiple imputations through chained equations using sequential regression trees as the conditional models. MICE-CART is applied for parameter usage reduction and this method is able to perform tuning while capturing complex relations of the data. Other local approach is GMCimpute [13] that is imputation that perform Gaussian mixture clustering. Even though GMCimpute is a local approach algorithm ,it is able to using more global correlation information. In this algorithm, by using Expected Maximization , the data is clustered into estimated missing values components of Gaussian mixtures. A single estimated missing value is computed from each component and then to obtain the final estimated missing values, the value is averaged. Local correlation information in the data is use for GMCimpute through the components mixture. The concept of multiple parallel estimations of missing values are utilized for collateral missing value imputation (CMVE) technique for improvement of final estimation. Compared to BPCA,KNN and LLSimpute accuracy in normalized RMS error(NMRSE) on many datasets, CMVE [15] has been able to produce better accuracy. This calculation are involving data sets ovarian cancer and yeast sporulation time series data. Ameliorative missing value imputation (AMVI) [16] apply simulation of Monte Carlo for determination of optimal number of reference gene, therefore AMVI is more advanced than CMVE .The time series expression profiles used in AMVI have been reported to exhibit strong dependency between observations. Adaptive Bicluster-Based Approach or ABBA is a missing value estimator for binary matrices. This algorithm has been made less complex for better understanding and usage, but the amount of parameters tweaking is also higher. [17] verified when the rate of missing values are higher than normal the performance of this algorithm is better than KNN.

2.3 Hybrid Approach

Local correlation between genes dominates and techniques such as LLSimpute or KNNimpute perform better compared to BPCA or SVDimpute especially for heterogeneous data sets. This statement prove that the performance of the imputation techniques is affected by correlation structure in the data. Missing value imputation approach such as BPCA or SVDimpute are preferred in term of global correlation and the accuracy of the predicted missing value increase as it will affect the conclusion to the case related to the dataset.

Nevertheless, there are still some hybrid approach for missing value imputation like HPM-MI[18] that significantly improved data quality by use of best imputation technique after quantitative analysis of eleven imputation approaches. HPM-MI combine the K-mean clustering with Multilayer Perceptron and use 11 different missing value imputation techniques and selects the best (compact) clusters among the results. K-means clustering is also used to validate class labels of given data.

Another notable approach is the GA+SVR[19] that reduce the error between the model output and the input. It is also to tackle the issue of difficulty in obtaining optimal value for the vigilance parameters. This approach apply Genetic Algorithm is used to optimize the selections of attributes and Support Vector Regression predicts the decision attributes value. Other hybrid approach is Fuzzy C-means+SVR+GA. This Hybrid use for overcome the issue of no theoretical basis for the selection of the cluster (c) number. It is also to tackle the issue of difficulty in obtaining optimal value for the vigilance parameters.

Knn + Neural Networks[20] is develop to overcome the issue of inadequate input value because before this most of the techniques discovered data as inputs and estimate an output class. This model fails if one or more inputs are missing. Consequently, they cannot be used for decision-making purposes if the data variables are not complete.

Fuzzy C-means based imputation + genetic algorithm[21] that based on inductance loop detector outputs. By utilizing the weekly similarity among data, the conventional vector-based data structure is firstly transformed into the matrix-based data pattern. Then, the GA is applied to optimize the membership functions and centroids in the FCM model.

MIGEC[22] is approach of imputation that create a method that minimizing the bias iteratively and less requirement on prior knowledge to the distributions that estimate missing values by multiple operations based on clustering. The non-missing data instances are separated into several clusters. Then, the imputed value is obtained after multiple calculations by utilizing the information entropy of the proximal category for each incomplete instance in terms of the similarity metric based on Gray System Theory (GST).

Multi-layered artificial immune system and genetic algorithm hybrid[23] is for effectively partial impute missing data in high-dimensional spaces. The multi-layered artificial immune system creates and stores antibodies that bind to and annihilate an antigen. The genetic algorithm optimises the learning process of a stimulated antibody. The evaluation of the imputation is performed using the RIPPER, k-nearest neighbour, naïve Bayes and logistic discriminant classifiers.

HHONC[24] hybrid approach that solve the issue of bias if there are systematic differences between the observed and the unobserved data by uses a higher correlation of input neurons for better fitting properties which often leads to a higher number of learning parameters (weights).Then It finds the closest k-neighbours from the unknown instance and returns the average value of the real-valued labels of the neighbours. The distance of the attribute is zero when the two instances miss the values of the same attribute, but when only one has a missing value, a maximal distance is assigned.

HPF[25] hybrid approach is develop to avoid intervals determined by different cluster information. The hybrid algorithm utilizes the global optimization capability of particle swarm to search the imputations of missing attributes and the cluster prototypes and incorporates the gradient-based FCM to calculate membership information, which is helpful in improving the clustering performance and the computational efficiency.

Other hybrid approach for imputation missing value is AR-ANN[26] that develop for imputing missing values and jointly handling the nonlinearity problem. AR model was used only for determining the input layer structure of ANN.

ANN-CBR[27] hybrid approach is developed to integrate artificial neural network with case-based reasoning.The hybrid algorithm utilizes ANN's strength that focus on feature extraction and pattern recognition and feature from CBR that identify same cases despites missing value.

FCM-MLP[28] is hybrid imputation approach that utilize fuzzy c-means and multilayer perceptron. Fuzzy c-means clustering is used as 1st stage and cluster 's center for each attribute is generated for complete datasets. For 2nd stage, MLP is utilized for network modelling.

PCA-NF+GA[29] is hybrid imputation approach that integrate principal component analysis and neural fuzzy with genetic algorithm. The data sets is orthogonalize by PCA to ensure the model is better trained then the compressed data is trained by neuro fuzzy model follow by GA to generate missing value and error between missing value and generated by GA is used as evaluation for comparison. If the error is not minimal, new missing values are generated by GA until error is minimized.

RF-AANN-GA[30] is hybrid based imputation that integrates Random forest and Auto-associative Neural Network Hybrid with genetic algorithm. For prediction of missing values in data sets, random forest are utilized the prediction of missing value is recorded. The prediction missing value by RF are used as limit for the search space for GA in AANN-GA system.

AANN-GA-RF[30] is hybrid imputation approach that combine AANN-GA system in cascade with Random Forest. Data are divided into four sets of training, validation, testing and experimental to generate artificially generate incomplete data sets then propagated through the AANN-GA for obtaining missing value. The imputed data and testing data are made to form complete datasets.

K-mean+MLP[31] is hybrid imputation approach that involving local learning and global approximation. In stage 1, missing value replaced by using k-mean algorithm then refine the resultant approximant values by using multilayer perceptron. The complete data set is trained by MLP with attribute having missing values as target variable one at a time.

ANN-NN[32] is develop based on observed hydrological data in Malaysia. It utilize the principle of Artificial Neural network combined with Nearest Neighbor. The proposed model is able to produce pattern of imputed data which closely resembles that of the observed values where peaks of the observed hydrographs are mostly attainable proves that the proposed model is able to capture daily maximum rainfall series.

DMI [33] is hybrid imputation approach that utilize decision tree and expectation maximization. (DMI) applies EMI on various horizontal segments (of a data set) where correlations among attributes are high. EM based imputation techniques rely on the correlations among the attributes in a data set

2.4 Knowledge Assisted Approach

This knowledge assisted approach is the missing values imputation process that have integration of external information or domain knowledge. This approach is significantly increase the imputation's accuracy using domain knowledge especially for data sets with small number of samples which are noisy or have high missing rate.

One of the example of knowledge assisted approach is the Projection Onto Convex Set (POCS) [34] that exploits the biological occurrence of synchronization loss and correlation information between genes and arrays, which is known as flexible set theoretic framework. POCS executes local least square regression to performs PCA imputation to capture array-wise correlation, capture gene-wise correlation

, and restricts the squared power of the expressions profiles for synchronization loss capturing. whether local or global correlation structure prevails in the,by using POCS the optimal solution can be achieved in the data. This is because final solution is always dominated by the smallest yet most reliable constraint set to satisfy larger yet less reliable constraint sets. Gene ontology (GO) has three independent ontologies that explain gene products in terms of associated biological processes (BP), cellular components, and molecular functions (MF) [35] and also known as a well-accepted standard for gene function categorization and GO increase the accuracy of the imputation as proven in experimental results where the proportion of annotated genes was large at higher rates of missing values. By combines histone acetylation information into KNNimpute and LLSimpute, Histone Acetylation Information Aided Imputation (HAIimpute) is developed. HAIimpute develop to increase of missing value estimation accuracy [36]. To form the pattern expressions, HAIimpute used the mean expression of genes from each of the clusters. By obtained by fitting a linear regression model between the gene and pattern expressions , the missing values are obtained. The final missing values estimation given by a convex combination of linear regression imputations and secondary imputation using both KNNimpute and LLSimpute. Indicated by improved correlation between imputed genes and original complete genes,it has also been proven that HAIimpute has consistently improved the KNNimpute or LLSimpute. The Table 1 below shows the grouping of imputation algorithms considered in this review.

Table 1: Missing Value Imputation Algorithm Categorized into Different Classes

Title	Algorithm	Year	Class	Reference	Remark
Missing value estimation methods for DNA microarrays	SVDImpute	2001	Global	5	Singular Vector Decomposition is used to obtain eigengenes that is a set of pattern of mutually orthogonal expression
A Bayesian missing value estimation method for gene expression profile data	BPCA	2003	Global	6	Estimate arbitrary missing variables is new in terms of statistical methodology.
Improving cluster-based missing value estimation of DNA microarray data	MICE-CART	2010	Local	12	combination of multiple imputation of classification and regression trees (CART) and chained equations (MICE)
Missing value imputation for gene expression data: computational techniques to recover missing data from available information	KNNimpute	2001	Local	8	uses pairwise information between the target gene with missing values and the k nearest reference genes to impute the missing values
Gaussian mixture clustering and imputation of microarray data	GMCimpute	2004	Local	13	perform Gaussian mixture clustering

IOP Conf. Series: Journal of Physics: Conf. Series **892** (2017) 012004 doi:10.1088/1742-6596/892/1/012004

Missing value estimation for DNA microarray gene expression data: local least squares imputation	LLSimpute	2005	Local	10	apply a multiple regression model
Sequential local least squares imputation estimating missing value of microarray data	SLLimpute	2008	Local	11	performs imputation by starting from the least missing gene with least missing rates
Collateral missing value imputation: a new robust missing value estimation algorithm for microarray data	CMVE	2005	Local	15	Apply the concept of multiple parallel estimations of missing values are utilized for collateral missing value imputation
Ameliorative missing value imputation for robust biological knowledge inference	AMVI	2008	Local	16	Apply simulation of Monte Carlo for determination of optimal number of reference gene
ABBA: adaptive bicluster-based approach to impute missing values in binary matrices	ABBA	2010	Local	17	missing value estimator for binary matrices
Hybrid prediction model with missing value imputation for medical data	HPM-MI	2015	Hybrid	18	Use the best imputation technique after quantitative analysis of eleven imputation approaches
A hybrid method for imputation of missing values using optimized fuzzy c-means with support vector regression and a genetic algorithm	GA+SVR	2013	Hybrid	19	tackle the issue of difficulty in obtaining optimal value for the vigilance parameters
A hybrid method for imputation of missing values using optimized fuzzy c-means with support vector regression and a genetic algorithm	Fuzzy C-means+SVR +GA	2013	Hybrid	19	use for overcome the issue of no theoretical basis for the selection of the cluster (c) number

A novel hybrid approach to estimating missing values in databases using k-nearest neighbors and neural networks	KNN+NN	2012	Hybrid	20	overcome the issue of inadequate input value because before this most of the techniques discovered data as inputs and estimate an output class
A hybrid approach to integrate fuzzy C-means based imputation method with genetic algorithm for missing traffic volume data estimation	FCM+GA	2015	Hybrid	21	based on inductance loop detector outputs
Missing data analyses: a hybrid multiple imputation algorithm using gray system theory and entropy based on clustering	MIGEC	2014	Hybrid	22	create a method that minimizing the bias iteratively and less requirement on prior knowledge to the distributions that estimate missing values by multiple operations based on clustering
Partial imputation of unseen records to improve classification using a hybrid multi-layered artificial immune system and genetic algorithm	MAIS+GA	2013	Hybrid	23	for effectively partial impute missing data in high-dimensional spaces
Missing Value Imputation Using Hybrid Higher Order Neural Classifier	HHONC	2014	Hybrid	24	solve the issue of bias if there are systematic differences between the observed and the unobserved data
A Global Clustering Approach Using Hybrid Optimization for Incomplete Data Based on Interval Reconstruction of Missing Value	HPF	2015	Hybrid	25	utilizes the global optimization capability of particle swarm
Imputation of Missing Values in Daily Wind Speed Data Using Hybrid AR-ANN Method	AR-ANN	2015	Hybrid	26	imputing missing values and jointly handling the nonlinearity problem
Imputation of missing values by integrating neural networks and case-based reasoning	ANN-CBR	2008	Hybrid	27	integrate artificial neural network with case-based reasoning

Hybrid model for data imputation: using fuzzy c means and multi layer perceptron	FCM-MLP	2014	hybrid	28	utilize fuzzy c-means and multilayer perceptron
Imputation of missing data using PCA, neuro-fuzzy and genetic algorithms	PCA-NF+GA	2008	Hybrid	29	integrate principal component analysis and neural fuzzy with genetic algorithm
Missing data imputation through the use of the Random Forest Algorithm	RF-AANN-GA	2009	Hybrid	30	integrates Random forest and Auto-associative Neural Network Hybrid with genetic algorithm
Missing data imputation through the use of the Random Forest Algorithm	AANN-GA-RF	2009	Hybrid	30	combine AANN-GA system in cascade with Random Forest
A novel soft computing hybrid for data imputation	K-mean+MLP	2011	Hybrid	31	involving local learning and global approximation
Imputation of time series data via Kohonen self organizing maps in the presence of missing data	ANN-NN	2008	Hybrid	32	utilize the principle of Artificial Neural network combined with Nearest Neighbor
A decision tree-based missing value imputation technique for data pre-processing	DMI	2011	Hybrid	33	Utilize decision tree and expectation maximization. (DMI) applies EMI on various horizontal segments (of a data set)
Microarray missing data imputation based on a set theoretic framework and biological knowledge	POCSimpute	2006	Knowledge	34	exploits the biological occurrence of synchronization loss and correlation information between genes and arrays
Improving missing value estimation in microarray data with gene ontology	GOimpute	2006	Knowledge	35	increase the accuracy of the imputation as proven in experimental results where the proportion of annotated genes was large at higher rates of missing values
Missing value imputation for microarray gene expression data using histone acetylation information	HAimpute	2008	knowledge	36	used the mean expression of genes from each of the clusters

3. Imputation Result Validation

One of the important step in demonstrate the reliability and accuracy of imputation algorithm is validation of imputation result. Firstly, Validation method can be categorized in two that is external and internal validation. Validation of imputation method is done by using internal validation that use only information in the data sets. Internal validation is done by compute performance indices of imputed data with original data [37]. External validation is done by accessing the imputation effect on analysis of subsequent biological. External validation apply the external knowledge rather than information's in the data sets.

3.1 Internal Validation

The most widely used for analyzing the accuracy of missing value imputation algorithm is done by computing normalized root mean square error (NMRSE). The accuracy of the missing value imputation algorithm is determined by the lower the NMRSE, the accuracy is higher [37]. The equation of NMRSE is as follows:

$$NMRSE = \sqrt{\frac{\sum_{i=1}^{m}\sum_{k=1}^{n}(g_{ik}-\tilde{g}_{ik})^2}{\sum_{i=1}^{m}\sum_{k=1}^{n}(g_{ik})^2}} \quad (1)$$

Where g_{ik} represent the *k*th experiment for dataset value g_i, and g and $\hat{g}$ represent the true value and imputed value respectively.

3.1 External Validation

The validation of missing value imputation algorithm by using external validation is by examined the effect of external information such as functional annotation of pathway information to downstream biological analysis. In microarrays experiment, the objective is to detect gene sets with same functional roles and when clustered, GO term that are significant are searched among gene of each cluster and one way to analyze the performance of missing value imputation algorithm is by checking for significant enrichment of GO term in gene cluster[37]. The calculation of the enrichment *P*-value for each GO term , *t* and for each cluster is as follows:

$$p = \sum_{i=k}^{\min(b,T)} \frac{\binom{t}{i}\binom{B-T}{b-i}}{\binom{B}{b}} \quad (2)$$

Where b=number of gene in cluster,
K=number of gene in cluster with GO term(t)
B= number of gene in the dataset

4. Limitation

Each algorithm has its own advantages and disadvantages, so does the datasets being used for each missing values technique. Several studies have shown that the performance of missing values imputation algorithms is significantly affected by factors such as correlation structure in the data, the missing data mechanism, the distribution of missing entries in the data, and the percentage of missing values in the data [2].Selecting the right algorithm may significantly boost the accuracy of the imputation results since there is no single imputation algorithm that performs the best in every situation. Global methods such as SVDimpute and BPCA perform better on data sets with low entropy, whereas local methods such as LLSimpute and KNNimpute perform better with high entropy data sets.

Another limitation is related to assumptions made on the distribution of missing values. Missing values in microarray data sets are normally assumed to be missing at random in most studies on missing values imputation, even though this is not a realistic assumption as missing values tend to arise in a systematic manner in practice. This pattern can significantly affect the imputation's performance and thus need to be properly considered in data analysis or algorithm design [12]. It also depends on the experimental conditions across the columns as variations in conditions may result in a non-random distribution of missing values in the data matrix. This is because each column in a microarray data matrix comes from different experiments.

Another limitation is testing and improving the model for multi-class imbalanced classification problems. Imbalanced classification problem arises due to imbalanced distribution of instances among multiple classes present in the data set. To further improve the imputation accuracy, feature selection and/or dimension reduction techniques can be applied to the training data; for example, principal component analysis methods can be applied before the support vector training to reduce the training time.

5. Conclusion

In large-scale experiment analyses and studies, missing values have become a frequent problem in data mining and this affects the result or conclusion being made. Since many analyses require complete data set, missing values imputation is an essential pre-processing step in data analysis. Instead of introducing new algorithms, variations based on existing imputation approaches are worthy contributions to systematic evaluation of existing algorithms. With so many different algorithms, evaluation of suitable parameters and operating platform has to be performed for optimal execution of the algorithm. This is to avoid overwhelming development of new missing value imputation algorithms before the full functional capability of previous existing ones has been fully explored.

References

[1] Batista, Gustavo EAPA, and Maria Carolina Monard. "A Study of K-Nearest Neighbour as an Imputation Method." HIS 87.251-260 (2002): 48.
[2] Gautam, Chandan, and Vadlamani Ravi. "Data imputation via evolutionary computation, clustering and a neural network." Neurocomputing 156 (2015): 134-142.
[3] Dempster, Arthur P., Nan M. Laird, and Donald B. Rubin. "Maximum likelihood from incomplete data via the EM algorithm." Journal of the royal statistical society. Series B (methodological) (1977): 1-38.
[4] Liew, Alan Wee-Chung, Ngai-Fong Law, and Hong Yan. "Missing value imputation for gene expression data: computational techniques to recover missing data from available information." Briefings in bioinformatics 12.5 (2011): 498-513.
[5] Oba S, Sato MA, Takemasa I, etal. A Bayesian missing value estimation method for gene expression profile data. Bioinformatics 2003;19(16):2088–96.
[6] Troyanskaya O, Cantor M, Sherlock G, et al. Missing value estimation methods for DNA microarrays. Bioinformatics 2001;17(6):520–5.
[7] Troyanskaya, Olga, et al. "Missing value estimation methods for DNA microarrays." Bioinformatics 17.6 (2001): 520-525.
[8] Liew, Alan Wee-Chung, Ngai-Fong Law, and Hong Yan. "Missing value imputation for gene expression data: computational techniques to recover missing data from available information." Briefings in bioinformatics 12.5 (2011): 498-513.
[9] Troyanskaya, Olga, et al. "Missing value estimation methods for DNA microarrays ." Bioinformatics 17.6 (2001): 520-525.
[10] Kim H, Golub GH, Park H. Missing value estimation for DNA microarray gene expression data: local least squares imputation. Bioinformatics 2005; 21: 187-198.
[11] Zhang X, Song X, Wang H, et al. Sequential local least squares imputation estimating missing value of microarray data. Comp Biol Med 2008;38:1112–20.

[12] Bras LP, Menezes JC. Improving cluster-based missing value estimation of DNA microarray data. Biomolecular Eng 2007; 24:273–82.

[13] Ouyang M, Welsh WJ, Georgopoulos P. Gaussian mixture clustering and imputation of microarray data. Bioinformatics 2004;20(6):917–23.

[14] Cai Z, Heydari M, Lin G. Iterated local least squares microarray missing value imputation. J Bioinform Comput Biol 2006;4(5):935–57.

[15] Sehgal MSB, Gondal I, Dooley LS. Collateral missing value imputation: a new robust missing value estimation algorithm for microarray data. Bioinformatics 2005; 21: 2417-2423.

[16] Sehgal MSB, Gondal I, Dooley LS, Coppel R. Ameliorative missing value imputation for robust biological knowledge inference. J Biomed Inform 2008; 41: 499-514.

[17] Colantonio A, Pietro RD, Ocello A, Verde NV. ABBA: adaptive bicluster-based approach to impute missing values in binary matrices. Proceedings of the 2010 ACM Symposium on Applied Computing; Sierre, Switzerland. 1774304: ACM; 2010. p. 1026-33.

[18] Purwar, Archana, and Sandeep Kumar Singh. "Hybrid prediction model with missing value imputation for medical data." Expert Systems with Applications 42.13 (2015): 5621-5631.

[19] Aydilek, Ibrahim Berkan, and Ahmet Arslan. "A hybrid method for imputation of missing values using optimized fuzzy c-means with support vector regression and a genetic algorithm." Information Sciences 233 (2013): 25-35.

[20] Aydilek, Ibrahim Berkan, and Ahmet Arslan. "A novel hybrid approach to estimating missing values in databases using k-nearest neighbors and neural networks." International Journal of Innovative Computing, Information and Control 7.8 (2012): 4705-4717.

[21] Tang, Jinjun, et al. "A hybrid approach to integrate fuzzy C-means based imputation method with genetic algorithm for missing traffic volume data estimation." Transportation Research Part C: Emerging Technologies 51 (2015): 29-40.

[22] Tian, Jing, et al. "Missing data analyses: a hybrid multiple imputation algorithm using gray system theory and entropy based on clustering."Applied intelligence 40.2 (2014): 376-388.

[23] Duma, Mlungisi, et al. "Partial imputation of unseen records to improve classification using a hybrid multi-layered artificial immune system and genetic algorithm." Applied Soft Computing 13.12 (2013): 4461-4480.

[24] Panigrahi, Lipismita, Kaberi Das, and Debahuti Mishra. "Missing Value Imputation Using Hybrid Higher Order Neural Classifier." Indian Journal of Science and Technology 7.12 (2014): 2007-2014.

[25] Zhang, Liyong, et al. "A Global Clustering Approach Using Hybrid Optimization for Incomplete Data Based on Interval Reconstruction of Missing Value." International Journal of Intelligent Systems (2015).

[26] Shukur, Osamah Basheer, and Muhammad Hisyam Lee. "Imputation of Missing Values in Daily Wind Speed Data Using Hybrid AR-ANN Method."Modern Applied Science 9.11 (2015): 1.

[27] Ennett, Colleen M., Monique Frize, and Robin C. Walker. "Imputation of missing values by integrating neural networks and case-based reasoning."Engineering in Medicine and Biology Society, 2008. EMBS 2008. 30th Annual International Conference of the IEEE. IEEE, 2008.

[28] Azim, Shambeel, and Suhas Aggarwal. "Hybrid model for data imputation: using fuzzy c means and multi layer perceptron." Advance Computing Conference (IACC), 2014 IEEE International. IEEE, 2014.

[29] Hlalele, Nthabiseng, Fulufhelo Nelwamondo, and Tshilidzi Marwala. "Imputation of missing data using PCA, neuro-fuzzy and genetic algorithms."Advances in neuro-information processing. Springer Berlin Heidelberg, 2008. 485-492.

[30] Pantanowitz, Adam, and Tshilidzi Marwala. "Missing data imputation through the use of the Random Forest Algorithm." Advances in Computational Intelligence. Springer Berlin Heidelberg, 2009. 53-62.

[31] Ankaiah, Narravula, and Vadlamani Ravi. "A novel soft computing hybrid for data imputation." Proceedings of the 7th international conference on data mining (DMIN). 2011.

[32] Malek, M. A., et al. "Imputation of time series data via Kohonen self organizing maps in the presence of missing data." Engineering and Technology 41 (2008): 501-506.
[33] Rahman, Geaur, and Zahidul Islam. "A decision tree-based missing value imputation technique for data pre-processing." Proceedings of the Ninth Australasian Data Mining Conference-Volume 121. Australian Computer Society, Inc., 2011.
[34] Gan, Xiangchao, Alan Wee-Chung Liew, and Hong Yan. "Microarray missing data imputation based on a set theoretic framework and biological knowledge." Nucleic Acids Research 34.5 (2006): 1608-1619.
[35] Tuikkala, Johannes, et al. "Improving missing value estimation in microarray data with gene ontology." Bioinformatics 22.5 (2006): 566-572.
[36] Xiang, Qian, et al. "Missing value imputation for microarray gene expression data using histone acetylation information." BMC bioinformatics 9.1 (2008): 252.
[37] Moorthy, Kohbalan, Mohd Saberi Mohamad, and Safaai Deris. "A review on missing value imputation algorithms for microarray gene expression data."Current Bioinformatics 9.1 (2014): 18-22.

Machining Parameters Optimization using Hybrid Firefly Algorithm and Particle Swarm Optimization

Nur Farahlina Johari[1], Azlan Mohd Zain[1,b], Noorfa Haszlinna Mustaffa[1,c] and Amirmudin Udin[2,d]

[1]Soft Computing Research Group, Faculty of Computing, Universiti Teknologi Malaysia, 81310 Skudai, Johor, Malaysia
[2]Faculty of Education, Universiti Teknologi Malaysia, 81310 Skudai, Johor, Malaysia

azlanmz@utm.my

Abstract. Firefly Algorithm (FA) is a metaheuristic algorithm that is inspired by the flashing behavior of fireflies and the phenomenon of bioluminescent communication and the algorithm is used to optimize the machining parameters (feed rate, depth of cut, and spindle speed) in this research. The algorithm is hybridized with Particle Swarm Optimization (PSO) to discover better solution in exploring the search space. Objective function of previous research is used to optimize the machining parameters in turning operation. The optimal machining cutting parameters estimated by FA that lead to a minimum surface roughness are validated using ANOVA test.

1. Introduction

The first known use of optimization is since year 1857. Optimization problem is a process of making something as functional and effective as possible. Basically, mathematical procedures are involved in the optimization problem to find the maximum or minimum function by choosing systematically integer variable value. Hence, optimization problem also can be defined as the problems that need a solution to find the best way from all feasible solutions. Optimization has been used in many areas as all the possible solutions are analyzed regarding their criteria. In my studies, some of the areas that used optimization are machining (we can find many machining cases use optimization in their work), clustering as stated in [1], robotic area done by [2], scheduling, networking and many more.

Optimization always present in day life since ancient times. Individual has to produce solutions that serve the source limits in order to deal with the restrictions of some problems. Basically, nature observations are being used to solve the problems. Gradually, knowledge will be developed based on this nature observations and past experiences. Then, the nature observations are being tested in daily life to find a better solution.

In machining, proper selections of process parameters are crucial to produce a high-quality product. Nowadays, surface finish of the products is one of the main concerns by manufacturing industries. They have to depend on manufacturing handbook to obtain optimal process parameters but results in increasing the manufacturing cost and low product quality [3]. Therefore, proper selection of process parameters are important to achieved high-quality product. Hence, optimization is needed to find the optimal parameters that satisfying the product quality.

Turning is one of traditional machining operation that is widely used in manufacturing industries. The operation is to cut metal apart from the workpiece. The primary method of turning is rotation which is to move metal against the cutting tool. It uses a single point cutting tool on lathe machine to cut the workpiece. One of common type of workpiece that has been widely used in turning operation is carbon steel.

Based on machinery handbook, carbon steel is steel that has properties made up mostly of the element carbon. Its principal is hardening element, determining the level of hardness or strength attainable by quenching. It raises tensile strength, hardness, resistance to wear and abrasion as the carbon content of steel is increased. It lowers ductility, toughness and machinability.

Most of manufacturing industries used carbon steels in their production. Carbon steels are known by their high production volume and good formability. It also has excellent weldability [4]. In order to achieve high quality of products, surface roughness of the workpiece need to be optimized due to high requirement of low cost applications in machining industries [5].

One of the most important requirements in machining process is surface roughness. It measures the technological quality of a product and describes the geometry of the machined surface and surface texture. The quality of a surface played important role in evaluating the productivity of machine tool and machined parts. To obtain better surface roughness value, a proper setting of cutting parameters is necessary [6]-[7]. Current method to select the cutting parameters that always been used is mainly depends on machinist expert and machining data handbook [8].

There are many factors that affect the value of R_a. The factors are machining parameters, cutting tool properties, cutting phenomena and workpiece properties but among all those factors, machining parameters are highly affecting the value of R_a [9].

This paper focused on machining parameters that highly affect the surface roughness value. The parameters that have to be optimised are feed rate (X_1), cutting speed (X_2) and spindle speed (X_3). The optimization process of machine parameters will help to increase companies' efficiencies and hence reduce the production costs. It also helps in the production process using new materials and may increase the productivity.

2. Metaheuristic Algorithm

Metaheuristic is an iterative generation process which guides a subordinate heuristic algorithm by combining intelligently different concepts for exploring and exploiting the search space while learning strategies are used to structure information to find efficiently near-optimal solutions [10].

In computer science, a computational method is designed to optimize the problem by iteratively trying to improve the solution of a problem. There are few algorithms that can search very large spaces of candidate solutions. However, metaheuristics do not guarantee an optimal solution is ever found.

[11] concluded that metaheuristic approaches obtained better quality solutions compare to heuristic approaches especially on hybrid techniques. So, metaheuristic is very suitable for solving nondeterministic polynomial-hard (NP-hard) optimization and complicated search problems.

This is proved by [12] that metaheuristics are designed to deal with complex optimization problems because of the other optimization techniques are not very efficient to solve the problems. Hence, these approaches were recognized as one of the most practical approaches for solving complex problems.

2.1. Fundamental Properties of Metaheuristics

Metaheuristics are designed to search optimal solution for a problem. [13] stated that there are some fundamentals properties of metaheuristics:

1. Metaheuristics are strategies that act as guidance for searching process.
2. The goal of metahueristics is to explore the search space efficiently to find optimal or near-optimal solutions.
3. The techniques used to solve problems are from simple local search procedures to complex learning process.
4. Metaheuristic algorithms are approximation solutions and usually non-deterministic.

Many different metaheuristics are in existence to solve optimization problems. Somehow, there are some new variants are continually being proposed. In last decade, the nature observation of swarm intelligence (SI) has become more popular among researchers to solve the optimization problems. The swarm intelligence is made up of a population of agents that interact with each other in their environment [14]. The inspirations of SI come from nature, especially the biological systems where the algorithms imitate the behaviour of swarms of social insects including ant colonies, bird flocking, animal herding, bacterial growth and fish schooling [15].

The use of SI approach is because of their robustness, flexibility, decentralized and self-organized characteristics [16]. Furthermore, SI is able to operate themselves independently without guidance from coordinator or external controller [17].

The most significant algorithms that have many contributions to the field are Genetic Algorithm (GA), Firefly Algorithm (FA), Particle Swarm Optimization (PSO), Levy Flight, Support Vector Machine (SVM), Glowworm Swarm Optimization (GSO) Cuckoo Search Algorithm (CS) and many more [18]-[20]. Firefly algorithm (FA) which was introduced by Yang (2008) is a nature-inspired metahuristic algorithm that inspires the flashing behavior of fireflies where the purpose of the flash is to act as a signal to attract other fireflies or preys. For computer science area, this algorithm has been applied in various problems such as optimization, clustering, image processing, feature selection and many more. The main purpose of this paper is to propose FA in optimizing the cutting parameters to solve the optimization problem in turning operation and hybrid it with Particle Swarm Optimization to find a better solution.

3. Firefly Algorithm

Fireflies are winged beetles or insects that produce light and blinking at night. [21] finds that there are strengths in firefly behavior to be applied in solving real life problems. So, he introduced Firefly Algorithm in 2008 inspired by flashing behavior of the fireflies for protection and attract mates or prey. There are three assumptions; fireflies are unisex and will be attracted to each other, brightness plays important role to attract others and fireflies will move randomly when their brightness are same [22].

3.1. The attractiveness of the firefly, I

The degree of brightness of firefly i and the distance r_{ij} between the firefly i and the firefly j demonstrates its attractiveness, I in the algorithm [22].

$$I(r) = I_s / r^2 \qquad (1)$$

As mentioned above, the brightness of fireflies are associated to objective function. So, the brightness of a firefly is chosen to reveal its recent position of its objective function f (x).

$$I_i = f(x_i) \qquad (2)$$

The brighter firefly will attract the less bright firefly due to its certain attractiveness value β regarding to their distances.

$$\beta(r) = \beta_0 e^{-\gamma r^2} \qquad (3)$$

where β_0 is the firefly attractiveness value at r = 0 and γ is the media light absorption coefficient.

3.1.1. The movement towards attractive firefly. The movement of a firefly i at position x_i moving to a brighter firefly j at position x_j is described by:

$$x_{i+1} = x_i + \beta_0 e^{-\gamma r^2 ij} \left(x_j - x_i\right) + \alpha \left(\text{rand-} \tfrac{1}{2}\right) \qquad (4)$$

where $\beta_0 e^{-\gamma r^2}\left(x_i - x_j\right)$ is due to the attraction of the firefly x_j; so if $\beta_0 = 0$ then it turns out to be a simple random movement. The attractiveness of firefly is compared with previous position. The algorithm will update the position of firefly to a higher attractiveness value; otherwise the firefly will remain in the current position. The termination criterion of the algorithm is based on its predefined number of iterations.

In the algorithm, objective function should be associated to the fireflies' brightness. Their attractiveness makes them capable to subdivide themselves into smaller groups and each subgroup swarm around the local models. Thus, FA is suitable for multimodal optimization problems [23]. FA also is used in clustering to find the centers of the clusters of the objects. The results obtained for FA method are compared to other methods such as Artificial Bee Colony (ABC) and Particle Swarm Optimization (PSO) [24].

FA also can be in dynamic environment. In previous research, FA is used for local search but direct movement and angle of peaks are been used for increasing the convergence speed. Results showed better performance and accuracy compare to PSO [25]. In other case, FA is proposed on dynamic environments based on Moving Peak Benchmark (MPB) Problem. Some modifications are done to the algorithm for better performance [26].

[27] used FA for scheduling jobs on grid computing to dynamically create an optimal schedule within minimum makespan and flowtime. Results showed that FA achieved less makespan time compared to Min-Min and Max-Min heuristics in scheduling problem.

[28] introduced FA to solve mixed variable structural optimization problems. The algorithm has been tested in six different structural optimization problem and results showed that FA is more efficient than PSO, GA, Simulated Annealing (SA) and HS methods. However, the randomization parameter must be reduced when using the FA as the optimization progress.

In engineering area, FA is proved to gain high profit [29]. The researchers introduced FA to solve optimal dispatch problem and results showed that FA gave optimal results that satisfying the constraints with high profit. FA also can be hybridized with other technique to improve its processing time because sometimes, FA itself requires a longer time to search for optimal values [30].

4. Particle Swarm Optimization

Particle Swarm Optimization is a population based stochastic optimization technique inspired by social behavior of bird flocking or fish schooling. The technique is developed by Dr. Russell Eberhart and Dr. James Kennedy in 1995.

The technique starts with initializing the population with random solutions. Then, the particles in PSO search for optimal solution by updating the generations by following the current optimum particles. Each particle in PSO can memorize their current best to search for the next generation. The value of current best will be stored in *pbest*. When the particles have compared their fitness with the overall population, the current best value will be stored in *gbest* (global best). Then, the velocity of each particle will be updated toward its *pbest*.

PSO is widely used by previous researchers to optimize the cutting parameters in machining. [31] used PSO and GA to optimize milling process parameters to maximize the profit rate. The algorithms are chose because of their better capability compare to other techniques.

PSO is used to find optimum value of R_a in milling operation. The result of PSO is compared with GA using previous research. This paper shows that GA takes longer time to get the optimum value compare to PSO [32]. PSO optimized the machining process parameters in high speed end milling operation. The result is compare with GA and SA and showed that PSO has high convergence speed than others [33].

5. Hybrid Firefly Algorithm and Particle Swarm Optimization

An attempt to hybrid FA and PSO is due to no memory capability in FA [25], [34] and [35]. Previous works show that PSO is one of the popular methods for a global search. So, the combination of FA and PSO will find a better solution to explore the search space by applying the capability of PSO in memorizes its previous best in determining the next possible solution [36].

The proposed algorithm began with FA implementation. After each firefly has move to more bright, the fitness of each firefly will be compared with its previous to obtain *pbest*. Then, FA will continue its process until the termination criteria met. Once the termination criteria are met, an optimal solution is obtained and the process end. Otherwise, the process will iterate again until the termination criteria are met.

The hybridization of FA and PSO is a way to overcome the memorizing capability of FA compare to PSO. In this research, we used the capability of PSO in memorizing the previous best location of the solutions to update the next value in the search space. The proposed algorithm will search better solutions by using the previous best location of firefly (*pBest*) in previous iteration to update the next value.

The current position of firefly x_i is a set of coordinates indicates a point in the search space. A firefly is representing a set of solution for the problem. The algorithm compares the attractiveness of the new firefly position with the old one. If the new position produces higher attractiveness value, the firefly is moved to the new position and the value is stored in *pBest*; otherwise the firefly will remain in the current position. The value of *pBest* will be compared on later iteration to determine optimal solutions.

The algorithm will keep finding the better positions of the fireflies by updating light intensity, *I* based on *pBest*. New points are selected by adjusting the randomization factor as a step size. There are two ways on terminating FA technique which are based on an arbitrary predefined number of iterations or predefined fitness value. In this research, we decided to use the predefined number of iterations as the termination criteria of the algorithm.

Once the maximum number of iteration is met, the algorithm will be terminated. The fitness values are calculated based on the optimal solution represented by the fireflies.

6. Optimization model of turning operation

As stated in the previous section, the operation considered in this research is turning operation. Metal cutting process has a complex nature whereby there are some objectives that tangled with each other. For example, by increasing the rate of tool wear may increase the unit production cost. Hence, it decides the quality of the product. So, it is very important to select the most suitable/optimal parameters of the machine tools in lathe machine when undergo the turning operation.

The model proposed by [37] is used as the objective function in this research. NC lathe machine is used to machine medium carbon steel (AISI 1045) with molybdenum High Speed Steel (HSS) turning tool. The model used to find the optimal cutting conditions which are feed rate, cutting speed and spindle speed. The performance measured is R_a.

There are 3 levels for each parameters in the operation. The levels are as in Fig. 1:

Fig. 1: Levels for cutting parameters of turning operation

Parameter	**Levels**		
	-1	0	+1
Feed rate, X_1 (mm/min)	0.5	0.8	1.0
Cutting speed, X_2 (m/min)	47.12	70.69	94.25
Spindle speed, X_3 (rpm)	1000	1500	2000

The objective function is described as in Equation 5:

Minimize R_a =

$$-0.74372 - 0.5245 * X_1 + 0.015759 * X_2 + 0.000734647 * X_3 - 0.00000171683 * X_2 * X_3 + 0.52133 * X_1\text{^}2 - 0.0000892502 * X_2\text{^}2 - 0.000000213667 * X_3\text{^}2 \quad (5)$$

Fig. 2 shows a list of 15 runs of experimental results done by [37]. The combinations of machining parameters are randomized according ti Box-Behnken design. The researchers used qualitest micro-trio-gloss (model No. 4430Nr-05643) to calculate R_a. Fig. 2 shows the design matrix of the experiment:

Fig. 2: Design matrix

Run	Feed rate, mm/rev (X1)	Depth of cut, mm (X2)	Spindle speed, rpm (X3)
1	-1	-1	0
2	0	0	0
3	-1	1	0
4	0	1	1
5	0	1	-1
6	0	0	0
7	1	0	1
8	-1	0	-1
9	0	0	0
10	1	1	0
11	-1	0	1
12	0	-1	1
13	0	-1	-1
14	1	-1	0
15	1	0	-1

Fig. 3 shows the experimental results done by [37]:

Fig. 3: Experimental results

Run	Feed rate, mm/rev (X1)	Depth of cut, mm (X2)	Spindle speed, rpm (X3)	Surface roughness, µm (R_a)
1	0.5	47.12	1500	0.18
2	0.8	70.69	1500	0.25
3	0.5	94.25	1500	0.2
4	0.8	94.25	2000	0.14
5	0.8	94.25	1000	0.211
6	0.8	70.69	1500	0.27
7	1	70.69	2000	0.3
8	0.5	70.69	1000	0.18
9	0.8	70.69	1500	0.27
10	1	94.25	1500	0.315
11	0.5	70.69	2000	0.16
12	0.8	47.12	2000	0.15
13	0.8	47.12	1000	0.14
14	1	47.12	1500	0.29
15	1	70.69	1000	0.33

Firefly Algorithm (FA) is applied to the turning problem. [38] who is founder of FA stated that the values of FA parameters which are α= 0.1-0.5, β= 0.7-1.0 and γ= 1.0 can be used for most of optimization problems. The higher number on n leads higher computation time. The parameters setting of FA parameters for this study are as Fig. 4:

Fig. 4: FA parameter setting

FA parameter	n	N_iteration	A	β	γ
Value	20	50	0.5	0.1	1.0

The range of cutting parameters of turning operation will be set based on [37] for FA technique. The optimization results of minimum R_a value obtained by FA can be seen in Fig. 5:

Fig. 5: FA single optimization (R_a)

Run	Feed rate, mm/rev (X1)	Depth of cut, mm (X2)	Spindle speed, rpm (X3)	Surface roughness, µm (R_a)	Time (seconds)
1	0.5506	47.1200	2000.0000	0.1228	0.3273
2	0.5517	47.1205	1901.2468	0.1406	0.1803
3	0.5391	55.0531	2000.0000	0.1477	0.1844
4	0.7542	93.9650	2000.0000	0.1420	0.1622
5	0.5517	78.1944	2000.0000	0.1583	0.1677
6	0.7386	47.2012	1999.9896	0.1508	0.2079
7	0.5575	53.0074	1000.7791	0.1406	0.1781
8	0.5459	49.2890	1999.4866	0.1307	0.1606
9	0.5838	85.6099	2000.0000	0.1434	0.2102
10	0.5739	75.3312	2000.0000	0.1636	0.2072
11	0.6404	47.1206	1999.9927	0.1314	0.2042
12	0.70544	94.25	2000.0000	0.1292	0.1862
13	0.5508	47.1201	1999.9944	0.1228	0.1967
14	0.7114	47.1203	2000.0000	0.1442	0.1738
15	0.5519	55.9778	2000.0000	0.1505	0.2008

Fig. 6 shows the Analysis of Variance (ANOVA) results of surface roughness. The analysis was carried out at significance level of α=0.05, for confidence level of 95%. The F-ratio indicates the significant of all cutting parameters to R_a while the p-value is the probability of significant contributions of each parameter and must be in the range of 0 to 1. If p-value < 0.05, it shows that the observed different within cutting parameters are significant. From Fig. 6, we can see that all parameters are significant towards surface roughness since all the p-values are lower than 0.05.

Fig. 6: ANOVA for FA

Cutting	df	SS	MS	F-ratio	p-value
X_1	2	0.0330	0.033	265.78	0.0001
X_2	2	0.0014	0.0014	11.26	0.0122
X_3	2	0.0015	0.0015	12.35	0.0098

df: degree of freedom; SS: sum of square; MS: mean of square

Based on previous researchers, the values of Particle Swarm Optimization parameters that usually used for most optimization problems are n = 20-100, N_iteration = 5-50, ϕ_1 = 0.1-0.9 and ϕ_2 = 0.1-0.9 [39] where ϕ_1 is weightage of particle best and ϕ_2 is weightage of global best. The parameters setting of PSO for this paper are n = 20, N_iteration = 50, ϕ_1 = 0.4 and ϕ_2 = 0.6. The parameters were set to give the best results possible based on the algorithm. The ranges of cutting parameters are set based on [37]. The optimization results of minimum R_a value obtained by PSO can be seen in Fig. 7:

Fig. 7: PSO optimization

Run	Feed rate, mm/rev (X1)	Depth of cut, mm (X2)	Spindle speed, rpm (X3)	Surface roughness, µm (R_a)	Time (seconds)
1	0.5577	51.6507	1998.7027	0.1393	0.1774
2	0.6193	51.7178	1998.7007	0.1450	0.2825
3	0.6248	51.7294	1998.6932	0.1457	0.2579
4	0.6783	51.6619	1998.7412	0.1538	0.3442
5	0.7223	51.7338	1998.6994	0.1631	0.3037
6	0.7321	51.6807	1998.6702	0.1652	0.1773

7	0.7012	51.6795	1998.6292	0.1583	0.3145
8	0.6119	51.6845	1998.5908	0.1441	0.1994
9	0.6433	51.6954	1998.6107	0.1482	0.2664
10	0.5678	51.6117	1998.5324	0.1399	0.2162
11	0.5750	51.5732	1998.5835	0.1402	0.2081
12	0.5272	51.4809	1998.5052	0.1376	0.2149
13	0.5974	51.5375	1998.4283	0.1421	0.2408
14	0.5864	51.5087	1998.4427	0.1410	0.2300
15	0.6519	51.4299	1998.4394	0.1487	0.2825

Fig. 8 shows the Analysis of Variance (ANOVA) results of surface roughness for PSO. The analysis was carried out at significance level of α=0.05, for confidence level of 95%. The F-ratio indicates the significant of all cutting parameters to R_a while the p-value is the probability of significant contributions of each parameter and must be in the range of 0 to 1. If p-value < 0.05, it shows that the observed different within cutting parameters are significant. From Fig. 8, we can see that the p-values for parameter X_2 and X_3 are lower than 0.05, means that the parameters are significant towards the R_a.

Fig. 8: ANOVA for PSO

Cutting	df	SS	MS	F-ratio	p-value
X_1	2	0.179	0.0898	3.1845	0.0593
X_2	2	0.189	0.0949	3.4176	0.0494
X_3	2	0.208	0.1041	3.8565	0.0353

df: degree of freedom; SS: sum of square; MS: mean of square

Fig. 9 shows the results of hybrid FA-PSO

Fig. 9: Hybrid FA-PSO optimization

Run	Feed rate, mm/rev (X1)	Depth of cut, mm (X2)	Spindle speed, rpm (X3)	Surface roughness, µm (R_a)	Time (seconds)
1	0.5601	47.1217	1878.6941	0.1445	0.1641
2	0.6337	49.9419	1047.9354	0.1431	0.1659
3	0.5644	55.5253	1000.0828	0.1519	0.3229
4	0.5726	50.2749	1000.0360	0.1283	0.1648
5	0.5004	49.5228	1920.7903	0.1454	0.2334
6	0.6014	94.2444	1918.6498	0.1344	0.1828
7	0.7460	83.3948	2000.0000	0.1769	0.1617
8	0.5915	47.1200	2000.0000	0.1257	0.1621
9	0.5656	47.1200	1750.1034	0.1605	0.1623
10	0.5756	85.4571	2000.0000	0.1432	0.1679
11	0.5030	47.1200	2000.0000	0.1216	0.1715
12	0.5117	47.1200	2000.0000	0.1216	0.1578
13	0.6079	87.0826	1979.9070	0.1465	0.2173
14	0.6170	69.4067	2000.0000	0.1713	0.1603
15	0.6319	94.2500	1749.1183	0.1738	0.1640

Fig. 10 shows the Analysis of Variance (ANOVA) results of surface roughness for hybrid FA-PSO. The analysis was carried out at significance level of α=0.05, for confidence level of 95%. The F-ratio indicates the significant of all cutting parameters to R_a while the p-value is the probability of significant contributions of each parameter and must be in the range of 0 to 1. If p-value < 0.05, it shows that the observed different within cutting parameters are significant. From Fig. 10, we can see that the p-value for parameter X_2 is lower than 0.05, means that the parameter is significant towards the R_a.

Fig. 10: ANOVA for hybrid FA and PSO

Cutting	df	SS	MS	F-ratio	p-value
X_1	2	1.0713	0.5356	1.2362	0.3083
X_2	2	4.1961	2.0980	6.9222	0.0042
X_3	2	0.8960	0.4480	1.0168	0.3768

df: degree of freedom; SS: sum of square; MS: mean of square

7. Summary

In this paper, a hybrid algorithm of FA and PSO is proposed and compared to single optimization of Firefly Algorithm and Particle Swarm Optimization to solve the problem in optimizing the cutting parameters for turning operation. Table 4 shows the comparison results of Firefly Algorithm, Particle Swarm Optimization and Hybrid FA-PSO:

Fig 11: Comparison results of FA, PSO and Hybrid FA-PSO

	Feed rate, mm/rev (X1)	Depth of cut, mm (X2)	Spindle speed, rpm (X3)	Surface roughness, µm (R_a)	Time (seconds)
FA	0.5506	47.1200	2000.0000	0.1228	0.3273
PSO	0.5272	51.4809	1998.5052	0.1376	0.2149
Hybrid FA-PSO	0.5117	47.1200	2000.0000	0.1216	0.1578

Based on the result, we can see that the proposed hybrid algorithm outperformed FA and PSO. The basic firefly algorithm is very efficient, but we can see that the solutions are still changing as the optima are approaching. It is possible to improve the solution quality by reducing the randomness gradually. However, due to no memory capability of firefly to memorize their movement, it takes some time to find the optimal solution. So, an element of PSO is used to cope with the problem. Furthermore, the hybrid algorithm can be modified to solve multi objective optimization problems.

References

[1] J. Lyko and R. Rudek, "An Algorithmic Analysis of Degressively Proportional Apportionments", *Journal of Computer Science & Computational Mathematics*; Vol. 6 Issue 3, p 53-57, 2016.

[2] R. Soto, S. Caro and B. Crawford, "On the Pursuit of Reliable Solutions for a Robotic Optimization Problem", *AASRI Conference on Intelligent Systems and Control*; Proceedia 4, p 26-30, 2013.

[3] A. Aggrawal and H. Singh, "Optimization of machining techniques – A retrospective and Literature Review", Sadhana, 2005; Vol 30, p 699-711, 2005.

[4] G. L. Huyett, "26th Edition Machinery Handbook", *Industrial Press, Inc. of New York, NY.* 2007.

[5] S. Thamizhmanii, S. Saparudin and S. Hasan, "Analyses of surface roughness by turning process using Taguchi method", *Journal of Achievements in Materials and Manufacturing Engineering.* Vol. 20 (1-2), p. 503-506, 2007.

[6] A. M. Zain, H. Haron and S. Sharif, "Review of ANN Technique for Modeling Surface Roughness Performance Measure in Machining Process", Proceddings – 2009 3rd Asia International Conference on Modelling and Simulation, AMS 2009.

[7] N. Yusup, A. Sarkheyli, A. M. Zain, S. Z. M. Hashim and N. Ithnin, "Estimation of Optimal Machining Control Parameters using Artificial Bee Colony", J Intell Manuf (2014) 25: 1463. doi:10.1007/s10845-013-0753-y, 2014.

[8] S.B. Raja, N.S. Narayanan & C. V.S. Pramod, A. Ragunathan, S.R. Vinesh, K.V. Krishna, "Optimization of Constrained Machining Parameters in Turning Operation Using Firefly Algorithm", Asian Network for Scientific Information, Journal of Applied Sciences, 2012.

[9] P. G. Benardos & G. C. Vosniakos, "Predicting surface roughness in machining: a review"

International Journal of Machine Tools and Manufacture, *43*(8), 833-844, 2003.
[10] I.H. Osma and G. Laporte, "Metaheuristics: A bibliography", *Annals of Operations Research.* Vol. 63, pp 513-623, 1996.
[11] L.Y. Tseng, "Metaheuristics Method and Their Applications.", 2003.
[12] S. Olafsson, "Metaheuristics. *Handbook on Simulation, Handbooks in Operation Research and Management Science,.* Volume 7, pp 633-654, 2006.
[13] C. Blum and A. Roli, "Metaheuristics in combinatorial optimization: Overview and conceptual comparison", *Journal in ACM Computing Surveys.* Vol. 35, Issue 3, pp 268-308, 2003.
[14] D.J.T. Sumpter, "The principles of collective animal behaviour", *Philosophical transactions of the Royal Society of London.* Vol 361, pp 5–22, 2006.
[15] C. Blum, and X. Li, "Swarm Intelligence in Optimization.", *Natural Computing Series. Springer-Verlag Berlin Heidelberg,* 43-85, 2008.
[16] E. Bonabeau, "Swarm Intelligence.", *O'reilly Emerging Technology Conference.* 2003.
[17] M. Dorigo and M. Birattari, "Swarm Intelligence.", *Scholarpedia.* 2(9):1462, 2003.
[18] A. M. Deris, A. M. Zain and R. Sallehuddin, "Hybrid GR_SVM for Prediction of Surface Roughness in Abrasive Water Jet Machining", Meccanica, 48: 1937, doi:10.1007/s11012-013-9710-2, 2013.
[19] N. Zainal, A. M. Zain, N. H. M. Radzi and A. Udin, "Glowworm Swarm Optimization (GSO) Algorithm for Optimization Problems: A State-of-the-Art Review", Applied Mechanics and Materials, Vol. 421, pp. 507-511, 2013.
[20] A. Mohamad, A. M. Zain, and N. E. N. Bazin, "Cuckoo Search Algorithm for Optimization Problems – A Literature Review", Applied Artificial Intelligence, An International Journal, Volume 28, Issue 5, 2014.
[21] X. S. Yang, "Nature-Inspired Metaheuristic Algorithms", *Luniver Press.* Second Edition, 2010.
[22] X.S. Yang, "Firefly Algorithm, Levy Flights and Global Optimization", *Research and Development in Intelligent Systems XXVI*, doi 10.1007/978-1-84882-983-1_15, 2010.
[23] X. S. Yang, "Firefly Algorithm, Stochastic Test Functions and Design Optimization", Int. J. Bio-Inspired Computation. Vol. 2, No. 2 (2010), pp.78–84, 2010.
[24] J. Senthilnath, S.N. Omkar and V. Mani, "Clustering using firefly algorithm: Performance study", Swarm and Evolutionary Computation. Vol. 1 (2011), p.164-171, 2011.
[25] S.H. Farahani,, B. Nasiri,and M.R. Meybodi, "A Multiswarm based Firefly Algorithm in Dynamic Environments", *In IEEE proceedings international symposium on information technology,* 2011.
[26] B Nasiri and M.R. Meybodi, "Speciation based firefly algorithm for optimization in dynamic environments", 2011.
[27] A. Yousif, A.H. Abdullah, S.M. Nor, and A.A Abdelaziz, "Scheduling Jobs on Grid Computing Using Firefly Algorithm", *Journal of Therotical and Applied Information Technology.* Vol. 33, No. 2, 2011.
[28] A.H. Gandomi, X.S. Yang, and A.H. Alavi, "Mixed variable structural optimization using Firefly Algorithm", *Computers and Structures.* Vol. 89, pp. 2325-2336, 2011.
[29] J. Ivanecky and D.H. Ing, "Optimal Dispatch of Renewable Energy Sources Including in Virtual Power Plant using the Firefly Algorithm", 2011.
[30] N.F. Johari, A.M. Zain, N.H. Mustaffa and A. Udin, "Firefly Algorithm for Optimization Problem", Applied Mechanics and Materials. Vol. 421 (2013), p. 512-517, 2013.
[31] U. Deepak, "Optimization of Milling Operation using Genetic and PSO Algorithm." *Bonfring International Journal of Software Engineering and Soft Computing*. Vol. 1, 2011.
[32] V. Pare, G. Agnihotri and C.M. Krishna, "Optimization of Cutting Conditions in End Milling Process with the Approach of Particle Swarm Optimization", *International Journal of Mechanical and Industrial Engineering (IJMIE)*. Vol. 1, Issue 2, 2011.
[33] F. Cus, U. Zuperl and V. Gecevska, "High speed end-milling optimization using Particle Swarm Intelligence", *Journal of Achievements in Materials and Manufacturing Engineering*. Vol.

22 Issue 2, 2007.

[34] S.K. Pal, C.S. Rai and A.P. Singh, "Comparative Study of Firefly Algorithm and Particle Swarm Optimization for Noisy Non-Linear Optimization Problems", *I.J. Intelligent Systems and Applications*. Vol. 10, pp 50-57, 2012.

[35] A. Hashmi,, N. Goel, S. Goel and D. Gupta, " Firefly Algorithm for Unconstrained Optimization", *IOSR Journal of Computer Engineering*. Vol. 11, pp 75-78, 2013.

[36] R. Poli, J. Kennedy, and T. Blackwell, " Particle swarm optimization", *Swarm Intelligence (2007), Springer*. Vol. 1, pp 33-57, 2007.

[37] S. Abdulkareem, U. J. Rumah and A. Adaokoma, "Optimizing Machining Parameters during Turning Process", International Journal of Integrated Engineering, Vol. 3 No. 1, p 23-27, 2011.

[38] X.S. Yang, S.S.S. Hosseini and A.H. Gandomi, "Firefly Algorithm for solving non-convex economic dispatch problems with valve loading effect", *Applied Soft Computing*. 12, 1180-1186, 2012.

[39] Y.Z. Lee, and S.G. Ponnambalam, "Optimisation of multipass turning operations using PSO and GA-AIS algorithms", *International Journal of Production Research*. Vol. 50, No. 22, pp 6499–6518, 2012.

IOP Conf. Series: Journal of Physics: Conf. Series **892** (2017) 012006 doi:10.1088/1742-6596/892/1/012006

Towards an Enhanced Aspect-based Contradiction Detection Approach for Online Review Content

Siti Nuradilah Azman[1], Iskandar Ishak[1], Nurfadhlina Mohd Sharef[1] and Fatimah Sidi[1]

[1]Faculty of Computer Science and Information Technology, Universiti Putra Malaysia, Serdang, Selangor, Malaysia.

iskandar_i@upm.edu.my

Abstract. User generated content as such online reviews plays an important role in customer's purchase decisions. Many works have focused on identifying satisfaction of the reviewer in social media through the study of sentiment analysis (SA) and opinion mining. The large amount of potential application and the increasing number of opinions expresses on the web results in researchers interest on sentiment analysis and opinion mining. However, due to the reviewer's idiosyncrasy, reviewer may have different preferences and point of view for a particular subject which in this case hotel reviews. There is still limited research that focuses on this contradiction detection in the perspective of tourism online review especially in numerical contradiction. Therefore, the aim of this paper to investigate the type of contradiction in online review which mainly focusing on hotel online review, to provide useful material on process or methods for identifying contradiction which mainly on the review itself and to determine opportunities for relevant future research for online review contradiction detection. We also proposed a model to detect numerical contradiction in user generated content for tourism industry.

1. Introduction

The ease of producing numerous user generated content online has led to a significant impact on the quality of information which is accessible to users worldwide. Currently, tourism field have widely used the facilities in web services as their one of the main source to understand the consumer's behaviours based on their reviews towards their products and services. The World Wide Web creates an important medium in information exchange between consumer and the producer[1]. These information between the consumer, industry suppliers which in this case hotels, attractions, intermediaries as such travel agencies, controllers such as governments and administrative bodies, as well as non- profit organizations plays an important role in travel sector revenue.

As stated by Xiang and partners [2], search engines, online travel booking websites and websites of marketing organization assist in information exchange between travelers. User generated content on the Internet has been particularly important in a travel sector[3]. At this age, user generated content such as online review has become a trend for every consumer when purchasing any services online. Consumer can simply write whatever they thought for a specific hotel or services that they purchased for other consumer to read. Take TripAdvisor.com for instance, is an example of a popular website that collects and published consumer reviews of hotels, restaurants and other travel-related services..

Primarily, online reviews could greatly improve consumers' ability to evaluate products. That is why online reviews are an important source of information for customers booking accommodation and travel [4]. Even if they do not book their travel online, most customers will at least review Online Travel Agencies (OTAs) and review sites before making a booking decision. However, the differences among users' preferences and perspective may lead to conflicting opinions about the target product, which could confuse a potential consumer when online reviews are one of the main sources of product information [5]. The increasing use of mobile devices for making travel bookings further enhances the importance of online reviews for travel decisions.

In order to observe the possible significance of this issue, we undertake a systematic literature analysis of the event. The paper reviews a number of papers in this field and presents the overview of identify contradiction in online review. The objectives of this paper are: (1) To investigate the type of contradiction in online reviews which mainly focusing on hotel online review, (2) To provide useful material on process or methods for identifying contradiction which mainly on the review itself, (3) To determine opportunities for relevant future research for online review contradiction detection.

This paper is structured as follows: Section 2 contains the contradiction issues in user generated content. In Section 3 contains the literature review on previous works regarding contradictory information in user generated content, and followed by Section 4 where we proposed a contradiction detection model on numerical mismatch. Lastly, Section 5 serves as the conclusion of this paper.

2. Contradictions in Online Review

The review strategy are based on the established methods of System Literature Review (SLR) guidelines for Software Engineering by [6] and Social Science [7]. Referring to the SLR guidelines, the systematic review is consists of 3 phases that includes planning the review, executing the review and formatting the report as shown in Figure 1.

The output of planning stage is review protocol which was structured on specifying research context, defining review protocol and structuring a research questions. This is then followed by the executing stage that performs search strategy and data extraction to categorize data items as the output. In reporting stage, the review will be documented in result and discussion.

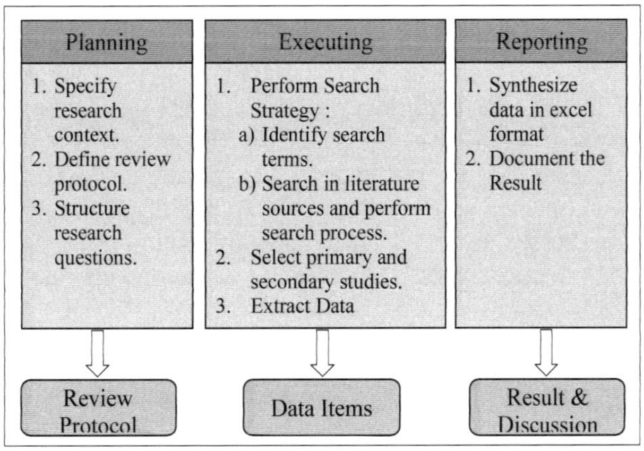

Figure 1. Systematic Review Phases

2.1. Definition

Online review is basically an opinion, view, sentiment or judgment of a consumer towards a business, product or services that is post online as such a review site. A consumer's review, is user-oriented, describing attributes in terms of usage situations, and measuring performance from a user's perspective[8]. Any individual can upload content online, in the form of images, audio or video for others to respond and comment, which forms the basis of Web 2.0. An online review is supposed to be helpful to other consumers as it plays a major role for travel purchase decisions.

The increasing use of mobile devices for making travel bookings further enhances the importance of online reviews for travel decisions. Customers trust reviews by other travellers more than they do official business communication, because they assume that the reviews are independent. Although some issues around fake reviews have reduced the trusts that consumers place in them, they still affect most travel decisions. However, the presence of conflicting or contradictions in online review, given a specific item or services, may have at least given negative effects on consumer and producer surplus.

There were several popular review sites that consumer can expose their innermost thoughts about a product. The first one would be Amazon.com. Amazon was one of the first online stores to allow consumers to post reviews of products in 1995, and it remains one of the most important resources for consumers looking to make informed purchase decisions.

Social media and consumer-generated content on the Internet continue to grow and impact the hospitality industry [9]. For Tourism and Hospitality domain, the most popular website is TripAdvisor.com [10]. It enables registered members to post reviews on hotels, restaurants, airlines, entertainment places to the site. We believe that online reviews play a vital role for consumer to purchase products as such in TripAdvisor. Therefore these online reviews should be precise when it comes to describing the important facts about a product, because other consumer wouldn't want to be.

Online reviews were usually in a text manner where it is written by the reviewer in sentences or a paragraph or an essay. There is also reviewer who reviewed product by posting it using videos or audio but for the purpose of the study; we focus on Text Online Reviews. Most or the review data that were posted online are in unstructured form. An unstructured documents usually have title and the content that is not organized in any structured manner which it will not have any attributes. Therefore, in order to identify whether a text online review is contradict or not, we need to recognize what are the features of contradictions and based on what feature does a text online review contradicts to each other.

2.2. Preliminary Study: TripAdvisor.com

We examined the reviews made by consumers focusing in particular on hotel reviews on TripAdvisor.com. This positive review on *My Hotel* in *Bukit Bintang, Malaysia* shows some conflicting information on the location of the hotel and *Puduraya Bus Station*. From the excerpts that we've taken in TripAdvisor.com, we highlighted the contradicting information that is based on numerical contradiction between these consumer reviews. (Refer Table 1).

Based on Table 1 also, we have found that the type of attribute of a text that is needed to be examined is either qualitative or quantitative. For the distance aspect, the consumers describe as short and near of walking with various time stated in their review. According to the observation made this feature of contradiction were discussed by [11] in their study. They categorized this as numeric mismatch which there exist contradiction when two sentences that contains numeric (dates or time) information are not similar to each other.

IOP Conf. Series: Journal of Physics: Conf. Series **892** (2017) 012006 doi:10.1088/1742-6596/892/1/012006

Figure 1. Excerpts of Reviews in TripAdvisor.com

Table 1. Information Analysis

	Review	Extracted Aspect	Extracted Aspect	Extracted Aspect	Extracted Aspect
1	Short walk to Puduraya Bus Station	Distance (short)	Time (null)	Location (Puduraya Bus Station)	Means of Transportation (walk)
2	Near to Puduraya Bus Station, 5-7 minutes of walking	Distance (near)	Time (5 to 7 Minutes)	Location (Puduraya Bus Station)	Means of Transportation (walk)
3	8-10 minutes of walking to Puduraya Bus Station	Distance (null)	Time (8-10)	Location (Puduraya Bus Station)	Means of Transportation (walk)

3. Related Works

3.1. Contradiction Detection

Based on the studies that have been identified, the contradiction analysis area is relatively a growing of research. There were no established common framework for describing and modeling the relevant problems. Though, there were some studies have made few steps towards this direction.

The study that is related to contradiction detection on text started since 2008 where it is being treated with recognizing textual entailment. Recognizing textual entailment (RTE) was introduced in 2005 has the objective of discovering whether some text (T) entails a hypothesis (H). In other words, given two text fragments, it requires to recognize whether one text can be inferred from the other. This entailment defines a directional relationship between pairs of text expressions[12].

This two-way RTE task requires systems to label each entailment pair as either "entailed" or "not entailed", regardless of the hypothesis (H) is a "true" fact. As for example; T entails H or T does not entail H. However in 2008, a three-way RTE task which was introduced. The concept of contradiction is established. the three-way RTE task requires to label each entailment pairs as either "entailed", "contradicted" or "unknown". As for example; T entails H or H contradicts T or unknown between H and T.

RTE has been established and fostered by PASCAL Recognizing Textual Entailment Challenge (RTE) in 2005[13]. Whereby PASCAL produced RTE1, RTE2, RTE3 datasets. Followed by RTE4, RTE5, RTE6 and RTE7 corpora was produced by NIST. This challenge is at the heart of many Natural Language Processing (NLP) tasks which includes Information Retrieval, Information Extraction, Question Answering, Machine Translation and other tasks to explore the linguistic expression. PASCAL and NiST

Generally, RTE model which is based on [13] includes preprocessing, enrichment, graph generation, alignment and inference (Refer figure 3).

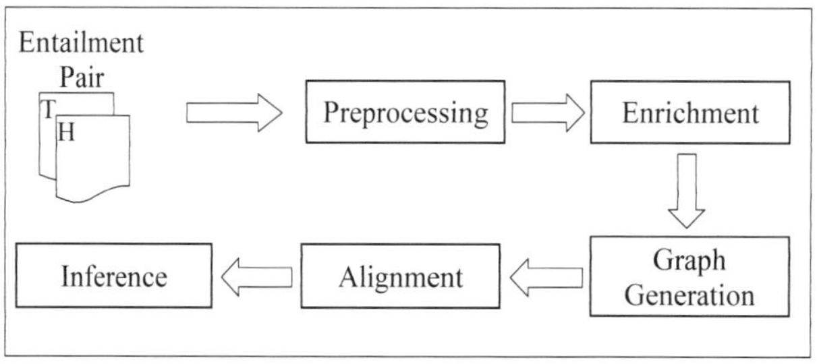

Figure 3. General Architecture of Textual Entailment. *Adapted from* [13].

Entailment pair needs to be pre-processed before launching a full-fledged analysis. Example of pre-processing annotators is part of speech tagging, dependency parsing, and some information extraction task like named entity recognition, reference resolution, semantic role labelling (SRL).

Enrichment is said to be differed with pre-processing by [13] as it involves either extraction or augmentation process of the output resulting from pre-processing. Extraction process is to abstract the annotation patterns by mapping them to its relation whereas augmentation is to enhancing the existing annotation by identifying implicit contents and changing it to explicit as new relation.

Graph generation is when the Text (T) and Hypothesis (H) are represented as graphs and being compared. Alignment step is basically the constituent or the output of analytics of T and H is being map. Alignment step is crucial in identifying the relevant portion of T and H which can simplify the inference step. Inference step is the part where decision on labelling the entailment pair. Number of approach has been introduced as for example discourse commitment extraction by [14], whereby implicit belief is extracted and decision-tree classifier is used to decide the entailment in inference step.

CD in text has also been treated with Sentiment Analysis (SA). SA is not a new direction of study; it has been a fundamental method for the task of detecting, extracting, and classifying opinions and sentiment which based on different topics that is expressed in textual input. SA is a part of area in text analytics and make use of techniques from natural language processing (NLP), information retrieval (IR), information extraction (IE) and artificial intelligence(AI). Generally, SA task includes (1) document selection (2) document processing; (3) document analysis (4) sentiment evaluation (Refer Figure 4).

IOP Conf. Series: Journal of Physics: Conf. Series **892** (2017) 012006 doi:10.1088/1742-6596/892/1/012006

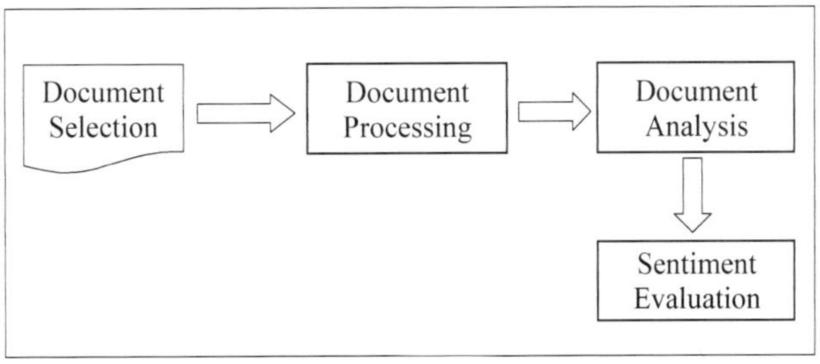

Figure 4. General Architecture of Sentiment Analysis. *Adapted from* [15].

At first step of sentiment analysis, the document will be selected and collected, based on the topic. Document processing generally involves information extraction (IE) techniques. Some of popular preprocessing technique is tokenization, stemming, parts of speech (POS) tagging, and feature extraction and representation. Tokenization is a technique in breaking down sentences into word, phrases or meaningful tokes by removing punctuation marks. Stemming is the process to convey a word into its root form. To recognize the different part of speech in text, parts of speech (POS) tagging is performed.

Document analysis step is where the classification technique of the sentiment analysis is being performed. According to [16], sentiment classification techniques consist of machine learning approach, lexicon based approach and hybrid approach. The Machine Learning Approach (ML) performs the famous ML algorithms and uses linguistic features. The Lexicon-based Approach relies on a sentiment lexicon, a precompiled sentiment terms. It is divided into dictionary-based approach and corpus-based approach which use statistical or semantic methods to find sentiment polarity. The hybrid approach combines both approaches and is very common with sentiment lexicons playing a key role in the majority of methods. In sentiment evaluation, the models are being evaluated to calculate the accuracy.

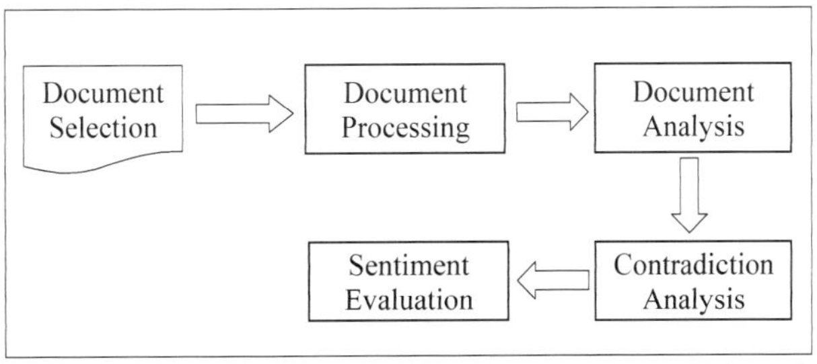

Figure 5. Sentiment Analysis approach for contradiction analysis. *Adapted from* [17].

Figure 5 shows additional contradiction analysis component is being added in SA. Study by [18] defines contradiction as sentiment diversity on document collections related to one or more topics and classify them based on time and the context. The author creates a novel contradiction measure based on mean value (μ^2) that indicates the aggregated sentiment and sentiment variance (σ^2) .

$$C = \frac{nM_2 - M_1^2}{\vartheta n^2 + M_2^2} \cdot W \qquad ....(1)$$

Where,

n	=	Cardinality of number of document
ϑ	$\neq$	0 ;used to limit the level of contradiction when μ^2 is close to zero
M_l	=	$\sum_{i=1}^{n} S_i$ - mean value (μ^2)

M_2 = $\sum_{i=1}^{n} S_i^2$ - sentiment variance (σ^2)
W = Weight function;
$\left(1 + \exp\left(\frac{\bar{n}-n}{\beta}\right)\right)^{-1}$ ....(2)
$\bar{n}$ = average number of topic document involved in the analysis;
β = scalling factor.

The sentiment *S* was defined as a real number in the range [−1, 1] that indicates the polarity of the author's opinion on Topic *T* expressed in a text. The aggregated sentiment (μ^2) was defined as the mean value over all individual sentiments assigned in that collection of documents *D* on topic *T*. The contradiction on a given topic *T*, between two groups of documents, $D_1, D_2 \subset D$ is defined in function of the information conveyed about *T*.

Based on these definitions, a three stages framework for contradiction detection was proposed. The first stage of the framework consists of identifying the topics for each sentence of the data. The second stage assigns a sentiment to each sentence-topic pair. Then, contradiction analysis is performed in the final step.

3.2. Contradiction Detection

Nonetheless, there were few features of contradiction that were put forth by previous research that is also can be examined (Refer Table 2). Based on our literature review, studies as such [14] put forth negation, antonymy and contrast. They define negations as directly or indirectly. Directly negations include (1) overt negative markers (e.g. don't, won't, can't) (2) negative quantifiers (e.g. no, no one, nothing) and (3) negative adverbs (e.g. never). Indirectly negations is categorized as (1) verb or phrasal verb (e.g. deny, refuse) (2) prepositions (e.g. without, except) (3) weak quantifiers (e.g. few, any, some) and (4) traditional negative polarity (e.g. anymore). Harabagiu et al. also stated antonymy as a feature of incompatibilities. Antonymy is when a word is directly the opposite of one another, as such "long or short" and "big or small". They detect antonymy by using WordNet lexicon that has more than 7000 antonymy relation. By analyzing sentences and phrases, they generate antonymy chains by using lexico-semantic chains. Contrast on the other hand is slightly differing from antonymy. Contrast is the course relation between separate phases or clauses within the same sentence. Contrast is detected when there is an occurrence of the word "but" or "although".

[11] introduces features of contradiction that consist of seven types of features for example 1) Number, date and time features; which this include the presence of numeric mismatch as mention earlier; 2) Structural features; syntactic structure of sentences (alignment of subjects), (example based on the subject: "Milo defeated Atticus" or "Atticus defeated Milo") ; 3) Polarity features; the presence of negative polarity text (example: "Not"); 4) Antonymy features; presence of aligned antonyms (example: "dark and bright) ; 5) Modality feature; presence of predefined modality markers (example: "can", "could", "may") 6) Factivity features; presence of factive construction (example: "forget", "remember", "know"); 7) Relational features; relations between sentences (example: "Corvus lives in Italy" and "Corvus would like to go to Italy").

Ambiguity is one of the features stated by [19]. It is a state where different classes of words, and the relationships between word as such homonyms and polysemy may create ambiguity. An expression is ambiguous if more than one meaning can be assigned to it. However, since this study is based on contradiction, this feature might not be included in the study.

Therefore based on this discussion for research question one, it is safe to say that, there exist features of contradiction in text online review which based on tourism industry. This direction of this study is to focus on numeric mismatch contradiction detection where the aspect of the review will be analyzed to identify contradiction in online review.

Table 2. Features of Contradictions

Features / Study	[11]	[14]	[20]	[19]	[21]	[22]
Negation		/			/	/
Antonymy	/	/		/	/	/
Contrast		/				
Number/Date/Time	/					
Structural	/		/			
Polarity	/		/			
Modality	/		/			
Factivity	/		/			
Relational Features	/		/			
Ambiguity				/		

4. Proposed Model

Limitation in identifying contradiction in text online review that we would like to highlight is based in the numerical feature of the text. As in Table 2, we highlighted the essence of numerical mismatch in online review. We also found that most of the study focuses on the data pre-processing and analysis in order to identify contradiction.

We found that the numerical feature detection only being discussed by [11] in detailed. Study by [11] proposed that all numerical aspects such as date and time expression will be normalized and numbers are represented by range. The aligned numbers is then considered contradict if the numbers is incompatible between T and H and the words are referring to the same entity.

Therefore, in this section we would like to propose an enhanced aspect-based contradiction detection in text for numerical mismatch. Study by [11] investigates the alignment scores of the T and H, we would like to suggest other ways on targeting to the problem such as hybridizing the aspect-based sentiment analysis and textual entailment general component(graph generation). The proposed contradiction detection focuses more on data processing and feature selection to achieve higher accuracy.

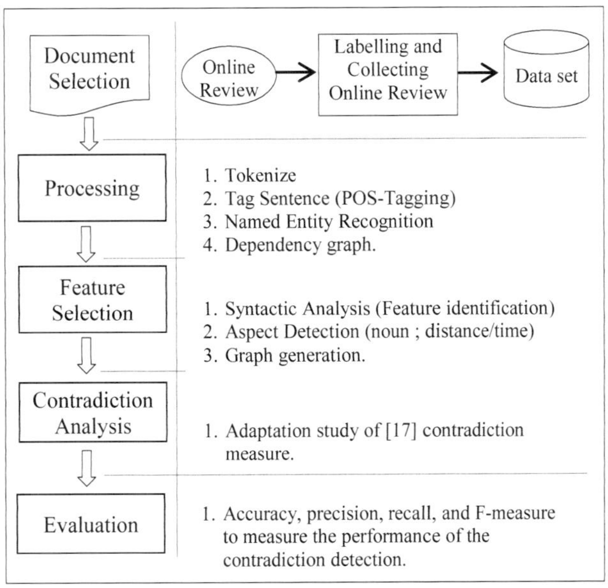

Figure 6: Proposed contradiction detection in text for numerical mismatch.

5. Conclusion

This paper investigates the type of contradiction in online review which mainly focusing on hotel online review, to provide useful material on process or methods for identifying contradiction which mainly on the review itself and to determine opportunities for relevant future research for online review contradiction detection. A total of 35 primary studies were selected and analyzed. Results of this review show that there was only one method for contradiction detection particularly in numerical contradiction. This paper also highlighted the issue of numerical contradiction that exists in user generated content for tourism industry. Also, the paper proposed a model for detection of contradictory information in online review for tourism industry. Our future work will be to verify our model to improve the accuracy of contradiction detection in online review which includes the numerical component such as distance and time.

References

[1] H. Werthner and S. Klein, "ICT and the Changing Landscape of Global Tourism Distribution," *Electron. Mark.*, vol. 9, no. 4, pp. 256–262, 1999.

[2] Zheng Xiang, K. Wober, and D. R. Fesenmaier, "Representation of the Online Tourism Domain in Search Engines," *J. Travel Res.*, vol. 47, no. 2, pp. 137–150, 2008.

[3] J. A. C. Dina Mayzlin, Yaniv Dover, "An Empirical Investigation of Online Review Manipulation," *Natl. Bur. Ecomomic Res.*, vol. 1, 2012.

[4] N. O. Mayer, "Online reputations; Why Hotel Reviews Matter and How Hotels Respond," no. June, 2015.

[5] H. Vali, D. Xu, and M. Yildirim, "The Effect of Conflicting Consumer Reviews on the Accuracy of a Purchase Decision," *AIS Electron. Libr.*, 2015.

[6] B. Kitchenham and S. Charters, "Guidelines for performing Systematic Literature reviews in Software Engineering Version 2.3," *Engineering*, vol. 45, no. 4ve, p. 1051, 2007.

[7] M. Petticrew and H. Roberts, *Systematic Reviews in the Social Sciences: A Practical Guide*. 2006.

[8] D. H. Park and I. Han, "Integrating conflicting reviews: Attributional hypotheses of consumer response to information uncertainty depending on prior brand attitude," *Proc. Annu. Hawaii Int. Conf. Syst. Sci.*, pp. 1–10, 2008.

[9] V. Browning, K. K. F. So, and B. Sparks, "The Influence of Online Reviews on Consumers' Attributions of Service Quality and Control for Service Standards in Hotels," *J. Travel Tour. Mark.*, vol. 30, no. 1–2, pp. 23–40, 2013.

[10] P. O'Connor, "User-Generated Content and Travel: A Case Study on Tripadvisor.Com," *Inf. Commun. Technol. Tour. 2008*, pp. 47–58, 2008.

[11] M. de Marneffe, an Rafferty, and C. Manning, "Finding Contradictions in Text," *Proc. ACL-08 HLT*, no. June, pp. 1039–1047, 2008.

[12] R. Shen, "Discovering Features Towards Recognising Textual Entailment," 2008.

[13] I. Dagan, D. Roth, and M. Sammons, *Recognizing textual entailment*. 2013.

[14] S. Harabagiu, a Hickl, and F. Lacatusu, "Negation, contrast and contradiction in text processing," *Proc. Natl. Conf. Artif. Intell.*, vol. 21, p. 755, 2006.

[15] R. Feldman, "Techniques and applications for sentiment analysis," *Commun. ACM*, vol. 56, no. 4, p. 82, 2013.

[16] W. Medhat, A. Hassan, and H. Korashy, "Sentiment analysis algorithms and applications: A survey," *Ain Shams Eng. J.*, vol. 5, no. 4, pp. 1093–1113, 2014.

[17] M. Tsytsarau and T. Palpanas, "Towards a Framework for Detecting and Managing Opinion Contradictions," *11th IEEE Int. Conf. Data Min. Work.*, pp. 1219–1222, 2011.

[18] M. Tsytsarau, T. Palpanas, and K. Denecke, "Scalable Discovery of Contradictions on the Web," *Proc. 19th Int. Conf. World Wide Web (WWW 2010)*, pp. 1195–1196, 2010.

[19] A. Ritter, D. Downey, S. Soderland, and O. Etzioni, "It ' s a Contradiction — No , it ' s Not : A

Case Study using Functional Relations," *Conf. Empir. Methods Nat. Lang. Process. (EMNLP 2008)*, no. October, pp. 11–20, 2008.

[20] M. Tsytsarau, T. Palpanas, and K. Denecke, "Scalable detection of sentiment-based contradictions," *CEUR Workshop Proc.*, vol. 762, pp. 9–16, 2011.

[21] C. Tu and M. Y. Day, "Chinese textual entailment with Wordnet semantic and dependency syntactic analysis," *Proc. 2013 IEEE 14th Int. Conf. Inf. Reuse Integr. IEEE IRI 2013*, pp. 69–74, 2013.

[22] T. L. Luyang Li, Bing Qin, "Generating triples based on dependency parsing for contradiction detection," *Commun. Comput. Inf. Sci.*, no. 4th National Conference, Social Media Processing, pp. 200–208, 2015.

IOP Conf. Series: Journal of Physics: Conf. Series **892** (2017) 012007 doi:10.1088/1742-6596/892/1/012007

An element search ant colony technique for solving virtual machine placement problem

J Srija[1], Rose Rani John[2] and Dr. Grace Mary Kanaga[3]

[1]Student, Karunya University, India
[2]Assistant Professor, Karunya University, India

e-mail : srijashilpa@gmail.com

Abstract. The data centres in the cloud environment play a key role in providing infrastructure for ubiquitous computing, pervasive computing, mobile computing etc. This computing technique tries to utilize the available resources in order to provide services. Hence maintaining the resource utilization without wastage of power consumption has become a challenging task for the researchers. In this paper we propose the direct guidance ant colony system for effective mapping of virtual machines to the physical machine with maximal resource utilization and minimal power consumption. The proposed algorithm has been compared with the existing ant colony approach which is involved in solving virtual machine placement problem and thus the proposed algorithm proves to provide better result than the existing technique.

1. Introduction

Cloud computing [1-3] is an innovative approach which provides computing services over the internet. The key technique behind cloud computing for designing, developing and delivering the services for the user, makes this approach ground breaking. Due to the attractive advantages like quality of service [ran and lamb], sustainability, accessing speed provided by the cloud environment, it has been widely accepted in short time. Any service provider has the motto to gain high profit simultaneously by reducing the operational cost [1, 2]. To achieve this, the concept of power consumption management has to be taken in to consideration One of the best ways to make cloud computing successful is through virtualization [7, 8]. Virtualization is a technique, which allows sharing a single physical instance of a resources or an application among multiple customers and organizations. It does by assigning a logical name to a physical storage and providing a pointer to that physical resource when demanded. The cloud service provider can ensure the quality of service by obtaining high resource utilization through virtualization.

Virtual machine placement is the process of effective mapping of virtual machines to the physical machines. The optimal mapping is necessary because it plays a major role in improving resource utilization and power maintenance [4-6]. Virtual machine placement is highly used in data center for cloud computing as a key technology to comprehend energy-efficient operation of servers . A data center usually contains thousands of blade servers which are tightly packed to increase the space utilization. The energy consumption problem has already attracted enough attention [9]. Many efforts have been made to improve the energy efficiency of data center from different aspects ncluding storage power management and network power management [10-14]. Numerous VMi tlacement schemes are designed and proposed for VM placement in the cloud computing environmenp aimed to improve various factors affecting the data centers, the VMs and their executions.

The next section states the related work which contains different approaches to solve virtual machine placement problem using different optimization algorithms. The third section contains the proposed work, general framework, algorithm explaining the working setup of the proposed system. The fourth section shows the result analysis and performance metrics which is then followed by conclusion and future scope.

2. Related Work

Dong et al. (2014) describe an Ant Colony Optimization based VM placement problem by considering multi-objective problem involved in each of the physical machines. However it is aimed to provide efficient network balancing thereby optimizing the total network traffic. The main focus is to achieve the modification in the availed traffic layout between the individual virtual machine. This modification is implemented by highly minimizing the whole of the traffic using the virtual machine placement in the data center and mapping the virtual machine with large traffic on the similar physical machine. The author also explains the concept of Maximum Link Utilization (MLU), through which network traffic is mapped evenly resulting in the avoidance of congestion. Moreover, for improving performance, they combine the ant colony optimization with 2-opt local search algorithm, because for large data calculation, ant colony optimization has a long way-finding, slow convergence and high time complexity.

Fei ma et al (2012) proposed a multi-objective ant colony optimization algorithm for reducing the SLA violation, resource wastage and power wastage in servers. In this proposed algorithm, for assigning any virtual machine to the physical machine each ant is made to perform solution construction. Once the solution is constructed, the evaluation process is performed, in which the pheromone level is increased only when the constructed solution appears in the list of best solutions, If the constructed solution doesn't appear in the best solution Otherwise, the pheromone level is made to get evaporated. At last the pheromone updation rule is done. By doing so, the best solution for virtual machine placement will be found.

Gao et al. (2013) present an ant colony system algorithm for analysing and finding solutions for virtual machine placement problem. This is obtained with the set of non-dominated solutions that has been developed for increasing resource utilization and reducing power consumption. The goal underlying this technique is that, complete utilization of the capacities of any servers will result in higher performance degradation. The efficiency is calculated and is compared with multi-objective genetic algorithm, two single-objective algorithms, a bin packing algorithm and a max–min ant system algorithm.

Wang et al. (2013) described the model of Virtual machine placement considering energy consumption of the data centers. The author proposed the method of optimizing energy consumption which is achieved by using PSO algorithm. The technique of redefining parameters provides an energy-aware local fitness first strategy. This method is highly efficient in updating the particle position thereby applying a two-dimensional particle encoding method which will improve the performance of PSO. This technique provides virtual machine replacement for energy consumption.

Zhen et al. (2015) proposed an algorithm for reducing the complexity in consolidated virtual machine placement. This algorithm proves to be novel and is based on biogeography method. This technique considers the consolidated placement problem as a complex system and tries to minimize both the resource wastage and the power usage. Moreover, the cpu and bandwidth are the important constraints considered for resource wastage and power consumption. This technique has been compared with the existing technique and finds to provide better performance

The problem underlying in multiple virtual machines which is to be mapped for the single request provided by the host is solved by Gupta et al. (2013). In this approach SLA requirements are considered as the main constraint for improving resource utilization. In order to achieve this, the High performance computing-aware scheduling algorithm has been designed for the VM placement which achieves better resource utilization.

Eagle algorithm has been proposed by Li et al (2011) To address the problem of the online VM placement in Li et al(2013). This algorithm is to reducing the total number of physical machines which are busy. In order to achieve this multi-dimension space partition model has been proposed. Whenever a new virtual machine arrives to get mapped with the physical machines, the resource utilization of each physical machine has to be calculated. Then the feasible physical machine will be selected to avoid unnecessary resource wastage. It is efficient in decreasing resource over the long run, and reducing the number of the PMs.

3. Proposed Methodology

The direct guidance ant colony system for virtual machine placement problem is aimed to provide effective mapping of virtual machines to the physical machines without wastage of resource and power consumption. The problem is formulated based on multi-dimensional space allocating algorithm and bin-packing algorithm. The proposed work contains two dimensional constraints namely the cpu utilization and memory The effective allocation is done based on selecting the physical machine effectively The physical machine is selected by analysing the utilized and idle capacity in the physical machine We use the direct guidance ant colony system for the effective selection of physical machine The resource wastage and power consumption model plays a major role for pheromone initialization and updation We also introduce the hash-set which contains the list of physical machines based on their capacity The hash-set has to be updated after every local-pheromone and global pheromone updation. Thus the necessity of the ant to search the entire datacentre for selecting the physical machine will be reduced by the usage of hash-set. This will in turn increase the computational speed

3.1 Resource wastage model

(Quingua Zhen et al 2012) formulated the resource utilization function based on the resource wastage model the resource utilization function has been developed to minimize the resource wastage efficiently.

$$W_j = | L^p{}_j - L_j{}^m | + \acute{\varepsilon} \ / U_j{}^p + U_j{}^m \quad (1)$$

Where,

W_j	- resource wastage
$L^p{}_i - L^p{}_j$	- normalized CPU and memory
$U_j{}^p + U_j{}^m$	- normalized remaining CPU and.
$\acute{\varepsilon}$	- Small positive real number 0.0001

3.2 Power consumption model

(F.Xang et al 2011) describes the power consumption modeling. The author clearly states that the power consumption is linearly proportional to the CPU utilization. For energy conservation servers are turned off when they are idle. Moreover, when the virtual machines are optimally allocated, the number of physical machine will highly reduce thus resulting in power consumption hence, their idle power is not part of the total energy consumption Based on above investigation, and power consumption (j) is introduced to quantify total power consumption by jth server. The formula stated by the author has been given below. This describes the power consumption factor and the parameters involved in the utilization of power and the possible way of maintaining the power utilization without any wastage. One should comprehend the power consuming factor in order to maintain the consumption.

$$P_j\,(j) = y_j{*}\,[\,(\,p_j^{\,busy} - p_j^{\,idle}\,) * (\,U_j^{\,P} + p_j^{\,idle}\,)] \qquad (2)$$

Where ,

$(\,p_j^{\,busy} - p_j^{\,idle})$ - Average power values indicating busy and idle power values.

3.3 Working of virtual machine placement using direct guidance ant colony

The ant colony approach is a biologically inspired approach which was developed by Dorigo in 1992. Based on the food-seeking behaviour of natural ants, the ant colony optimization came into existence. This optimization technique found successful and was utilized in various fields like genetic engineering, networking and so on. The description about the working setup of direct guidance ant colony system for virtual machine placement problem is given below:

3.3.1 Solution construction: The pheromones trail initialization is based on the chemical substance secreted by the ants while they move in search of food. This pheromone paves way for the new set of ants to find the path. In this paper, the pheromone trail initialization is calculated based on the result of power and resource wastage modeling. The pheromone trails are initialized on each of the physical machine by which resource and power utilization on each physical machines are calculated. In order to calculate the resource and power consumption modeling the two constraints cpu and memory should be considered. The cpu and memory demand of each virtual machines are calculated followed by the resource capacity of the physical machine in to which the virtual machines are to be allocated. Once they are calculated the physical machine containing the least capacity will get the highest pheromone value of 0.3 and the physical machine with higher capacity contains least initial pheromone value. Thus the initial trail value is set, as stated by [1], which is formulated as,

$$\tau_0 = 1 / [\, n{*}\,(P(s_0\,)) + (W(s_0\,))\,] \qquad (3)$$

Where

$$P(s_0\,) \;=\; \Sigma_{j=1}^{\;m}\;[Pj * P_j^{max}\,] \qquad (4)$$

Where,

P_j^{max} - Peak power consumption in j^{th} server
n - Number of virtual machines.

3.3.2 Local pheromone updation: Once the initial physical machine has been selected by the ant based on the availability of the pheromone trail, then the arrived virtual machines are allocated in to that physical machine. Once the allocation has been done then the local pheromone updation is performed. The local pheromone updation [1-2] is based on the formula (7).

3.3.3 Hash-Set updation: Then after the first iteration, the empty hash set should be updated. The hash-set is the one which contains the list of next physical machines which is to be selected. The hash-set is updated by the second group of ants which is initiated after the first local updation. This second group of ants is involved only in hash-set updation. Immediately after the first local updation, this second group of ants starts updating the hash-set. The hash-set updation is performed based on the below stated formula:

$$hs(i) = \tau_{i,j} + \rho_n \quad (5)$$

Where

hs(i) - hash set i
$\tau_{i,j}$ - pheromone value after current local updation
ρ_n - Pheromone decay value during that iteration

Once the hash-set updation is being done, in order to make it efficient the max and min trail limits are set for the physical machines capacity. Thus the physical machine with higher trail intensity is set as τ_{max} and the physical machine with least trail intensity is set as τ_{min}. The physical machines with intermediate capacities are sorted between τ_{max} and τ_{min}. This causes the hash-set to be filled with feasible set of physical machine, which is to be selected in which order.

3.3.4 Global Pheromone Updation: Thus if all the ants have completed the tour, global pheromone updation will be performed. After global pheromone updation, the global threshold value for each physical machine is set, in order to make the physical machine efficient for next iteration.

3.4 Algorithm

Step 1: Initialize the number of ants.
Step 2: Create the empty hash list set
Step 3: Initialize the trail intensity τ_0 using (1).
Step 4: Select the Virtual machine which has to be mapped in the host based on the pseudo-random proportional rule:

$$\rho_{i,j} = [\alpha * \tau_{i,j} + (1-\alpha)\, \acute{\eta}_{i,j}] \,/\, \Sigma_{i=1}{}^{n} (\, \alpha * \tau_{i,j} + (1-\alpha)\, \acute{\eta}_{i,j})] \quad (6)$$

Step 5: Perform local updation rule between the virtual machine i and host j

$$\tau_{i,j} = (1-\rho_l)\, \tau_{i,j}\, (t-1) + \rho_l * \tau_0 \quad (7)$$

Where,

τ_0 = initial pheromone factor
ρ_l = local pheromone evaporating factor whose value lies between 0.1-0.7.

Step 6: Initialize the second group of ants.

IOP Conf. Series: Journal of Physics: Conf. Series **892** (2017) 012007 doi:10.1088/1742-6596/892/1/012007

Step 7: Update the hash-set using the formula (5).
Step 8: Set max and min trail limit within the hash-set.
Step 9: Repeat this procedure until no virtual machines are available to be allocated to the server.
Step 10: For each set solution perform global pheromone updation

$$\tau_{i,j}(t) = (1-\rho_g)\, \tau_{i,j}(t-1) + [(\rho_g * \lambda) / (P(s) + W(s)) \quad (8)$$

Where,

$$\lambda = [\, NI / t + 1 \,] \quad (9)$$

NI_s = Number of iteration

Step 11: End for
Until maximum number of iterations are reached.

4. Work flow

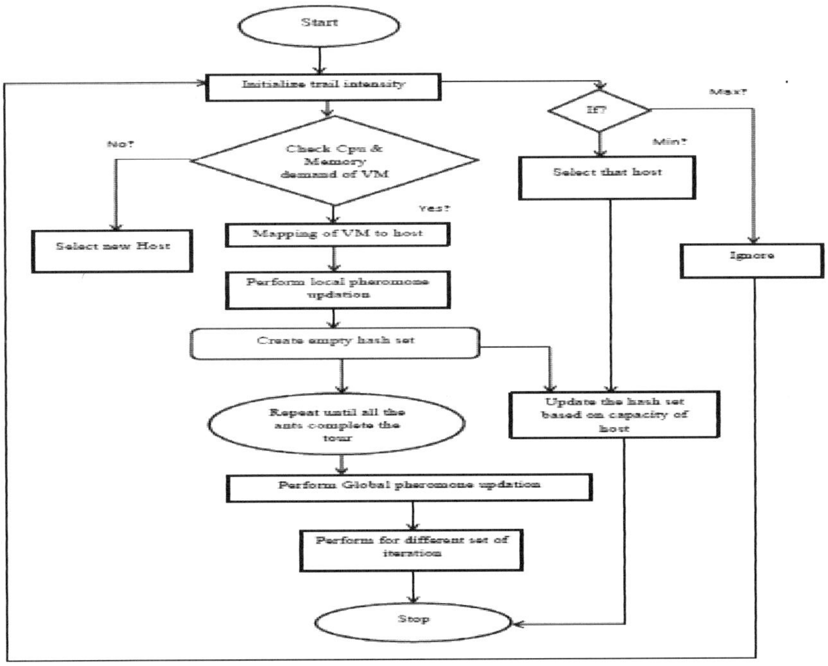

Figure 1. Work flow of direct guidance vmpp

5. Experimental Setup

The direct guidance ant colony system for virtual machine placement problem has been implemented using cloudsim. The datacentre has been created which contains the set of physical machine, virtual machine and the constraints associated with it. The constraints considered are cpu utilization and memory.

The main objective is to optimally allocate the virtual machine into the physical machine without the loss of energy and power. To achieve this direct guidance ant colony system has been proposed, in which the physical machine selection and the pheromone updation plays a major role for optimal mapping. Here the number of physical machines is set 45 and the virtual machines are modified from the range of 60-150. The threshold value of cpu utilization and memory utilization has been fixed to 0.6. As the virtual machines are varied, the number of physical machines taken for optimal mapping is calculated and is compared with the basic ant colony system and the results are tabulated. The computational time for allocating the virtual machines to the physical machines are analyze and compared with the basic ant colony algorithm for virtual machine placement problem and the resulted are tabulated.

6. Result Analysis

The direct guidance ant colony system for virtual machine placement problem has been implemented using cloudsim. The datacentre setup has been made which containing the set of physical machines that are busy and idle. The capacity of each physical machine based on cpu utilization and memory has been analysed. The set virtual machines with various memory and cpu demand that are to be mapped within the physical machines are selected using the above algorithm. Once any virtual machine is mapped, the local updation will be made at the physical machine and this process continues till the entire ant completes their process. Threshold value for cpu utilization and memory will be updated automatically, after every local updation. This will ensure that the virtual machine with higher demand of cpu utilization and memory than the physical machine not to be selected. Then the global updation with global pheromone decay factor is made to perform and the process repeats. This approach is tested with the basic ant colony system for virtual machine placement problem and this found to provide better result.

The graph has been plotted in order to indicate the effective virtual machine allocation in to the physical machine. This is calculated based on the above stated algorithm, in which the pheromone trail initialization, selection of physical machines, threshold value setup, pheromone updation and evaporation are explained clearly for getting the optimal mapping of virtual machine in to the physical machine.

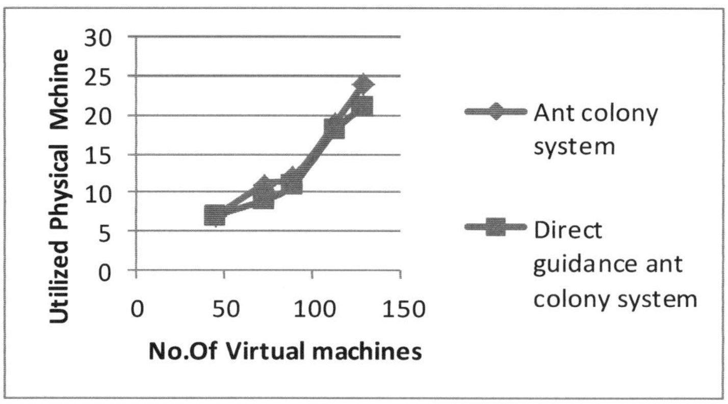

Figure 2. Mapping of virtual machines

After any iteration, the hash list containing the list of next physical machine to be selected is searched. The physical machines are selected based on their capacity which will be updation after single iteration. By doing so, the ant need not search for the best physical machine for which the virtual machine has to be placed, instead the physical machine based on their capacity are placed with max and min values could be selected easily.

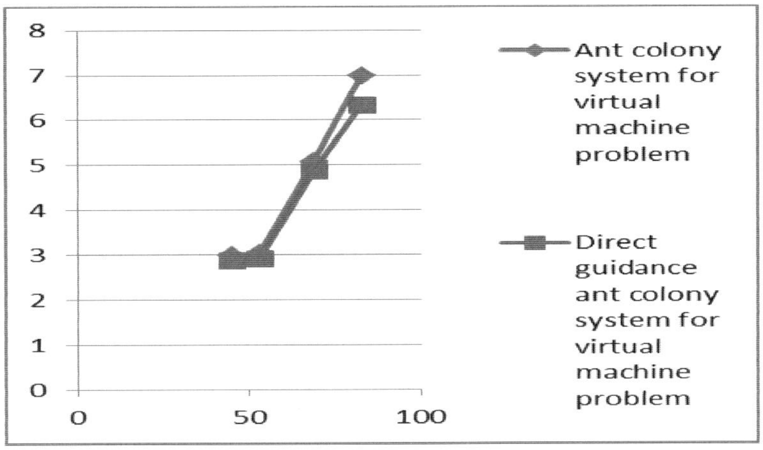

Figure 3. Computational time of vmpp

7. Conclusion

The main motive of direct guidance ant colony system for virtual machine is to attain the optimal solution with high computational speed thereby efficiently allocating the virtual machine to the host effectively which in turn will reduce the total number of host in the system and consumes energy. The direct guidance ant colony system involves the way of initializing the pheromone trail which avoids the random exploration of all the nodes, and the candidate list are updated effectively by segregating nodes based on their trail intrensity and hence the ants can easily select the feasible node for their next traversal. Thus the occurrence of non-feasible nodes could be highly avoided. Graphs are plotted in order to compare performance of the existing algorithm and the proposed algorithm, in which the proposed algorithm provides higher efficiency. This algorithm is implemented in an offline instance of Virtual Machine placement problem and the future scope is to enhance the direct guidance ant colony algorithm to implement it in the online instance of VM placement problems.

References

[1] Akhiro Uchido. "An Efficient GPU implantation of ant colony for travelling salesmen problem", Third International conference on Network and Computing, Scientific world journal volume, article ID 646953, 13 pages, 2013.
[2] Bifan Li, Lipo Wang, Wu-Sang. "Ant colony optimization for travelling salesmen problem for ants with memory", Fourth National conference on Natural Computation – Prospects and challenge (ICEE), Procedia-Natural computation 437-447, 2008
[3] Casper Jose Eyckelhef. Ant system on dynamic travelling salesmen problem", National conference on swarm intelligence 34, 719-729, 2002.
[4] Chi-Bin Cheng, Chun-Pin Maoi, "A modified ant colony optimization for travelling salesmen problem with time windows", Elsevier publication on Evolutionary computing – 1497-1517, 1996.
[5] Hozefa M. Botee, Eric Bonabeau, Sanjay Chaudary. "Evolving ant colony optimization in travelling salesmen problem", Computation and Research challenges, 1997.

IOP Conf. Series: Journal of Physics: Conf. Series **892** (2017) 012007 doi:10.1088/1742-6596/892/1/012007

[6] Jei Bai, "A model induced max-min ant colony optimization for travelling salesmen problem", Applied soft computing vol. 1, 2013.
[7] Jinhui Yang, Xiaohu Shi, Maurizio Marchese and Yanchun Liang, "Ant colony optimization for generalized travelling salesmen problem", Elsevier publication on Natural sciences, 2009
[8] Kan Jun Man, Zhang Yi, "Application of improved ant colony optimization algorithm for genetic engineering approach", International conference on future electrical power and energy system. 2015
[9] Krishna Hingrajiya, "An approach for solving multiple travelling salesmen problem using ant colony optimization", Fourth International conference on Computer engineering and Intelligence system, 1133-1149, 2015
[10] Rakib Hasan, Hashem, "Improved ACS for larger travelling salesmen problem", IEEE publication on Evolutionary computing vol.1 no.2, 2013.
[11] Thomas Stutzle and Dorigo,"Solving travelling salesmen problem using elitist ant system", International journal on swarm intelligence 54,2787-2805, 2016
[12] Xio Min-hue and Dorigo, "Solving travelling salesmen problem using Orthogonal ant system" , 2016.
[13] Yue-min Yue. "An improved ant colony optimization for travelling salesmen problem", International Journal on Evolutionary Computing, 2008.
[14] Zar Chi Su Hliang, May aye Khine. "Solving travelling salesmen problem by using improved ant colony optimization", International journal on Information and Education technology 129-156, 2016.

IOP Conf. Series: Journal of Physics: Conf. Series **892** (2017) 012008 doi:10.1088/1742-6596/892/1/012008

Association Rule-based Predictive Model for Machine Failure in Industrial Internet of Things

Jung-Hyok Kwon[1], Sol-Bee Lee[1], Jaehoon Park[2] and Eui-Jik Kim[1*]

[1] Department of Convergence Software, Hallym University, 1 Hallymdaehak-gil, Chuncheon-si, Gangwon-do 24252, South Korea
[2] Department of Electronic Engineering, Hallym University, 1 Hallymdaehak-gil, Chuncheon-si, Gangwon-do 24252, South Korea

ejkim32@hallym.ac.kr

Abstract. This paper proposes an association rule-based predictive model for machine failure in industrial Internet of things (IIoT), which can accurately predict the machine failure in real manufacturing environment by investigating the relationship between the cause and type of machine failure. To develop the predictive model, we consider three major steps: 1) binarization, 2) rule creation, 3) visualization. The binarization step translates item values in a dataset into one or zero, then the rule creation step creates association rules as IF-THEN structures using the Lattice model and Apriori algorithm. Finally, the created rules are visualized in various ways for users' understanding. An experimental implementation was conducted using R Studio version 3.3.2. The results show that the proposed predictive model realistically predicts machine failure based on association rules.

1. Introduction

Industrial Internet of things (IIoT) is a key enabler for realizing the fourth wave of the industrial revolution, since it can greatly improve manufacturing enterprises' operational efficiencies, such as performance optimization, safety improvement, time and cost savings, etc. [1–3]. IIoT generally employs many machines, and generates a huge amount of data by detecting and measuring changes in their surrounding environments. This data can be processed and analyzed to provide useful information for end users to manage the machines efficiently. In recent years, many manufacturing enterprises and research institutes have investigated big data analytics to extract meaningful values from the collected data [4].

The prediction of machine failure is one of the most challenging issues faced by manufacturing enterprises, since machine failures may seriously damage the manufacturing process and increase maintenance costs. Big data analytics can provide valuable insight into failure causes, and thus it is widely used to predict machine failure. The types and causes of machine failure vary widely. Therefore, for accurate prediction, it is very important to create rules based on relationships between the cause and type of machine failure. However, most existing predictive models for machine failure only focus on a single type of machine failure [5]. Therefore, these models are difficult to be applied to real enterprise manufacturing processes.

In this paper, we propose an association rule-based predictive model for machine failure in IIoT, which provides more realistic prediction results by considering various machine failure types. The predictive model was developed based on association rules considering relationships between failure causes and types. We considered three major steps: 1) binarization, 2) rule creation, 3) visualization. The first step creates a binary representation of item values. The second step creates the association rules, represented as an IF-THEN structure. The third step visualizes the rules in various ways for users' understanding.

To verify the feasibility of the proposed predictive model, we conducted an experimental implementation using R Studio version 3.3.2 [6]. As the dataset, we use the log data of 1,000 machines. The data for each machine includes four items for the cause of machine failure (i.e., CPU usage ratio, memory usage ratio, machine temperature, and low battery) and three items for the types of machine failure (i.e., long latency, shutdown, and freezing). The results show that the proposed predictive model provides the realistic prediction results based on the association rules.

The rest of this paper is organized as follows. In Section 2, the predictive model for machine failure is described in detail. The description for dataset and the implementation results are presented in Section 3. We conclude this paper in Section 4.

2. Predictive model for machine failure

To develop the machine failure prediction model, we used association analysis to create association rules identifying how the items are related [7]. The predictive model for machine failure is represented in the form of an IF-THEN structure based on the created association rules. The model development incorporated three major steps, as follows.

2.1. Binarization

The value of categorical items and continuous items were transformed into binary representation. Each item value was represented as 1 when it was present, and 0 otherwise. Thus, the continuous values were converted to categorical items by dividing into multiple categories, i.e., discretization. The number of categories and the range of each category can be predetermined. Figure 1 shows an example of binarization, where the number of categories for CPU usage ratio was set to 3: [0, 50], (50, 80], and (80, 100].

Machine ID	CPU usage ratio	Residual energy
1	22%	No
2	58%	Yes
…	…	…
N	82%	Yes

Binarization

Machine ID	CPU usage ratio = [0, 50]	CPU usage ratio = (50, 80]	CPU usage ratio = (80, 100]	Residual energy = Yes	Residual energy = No
1	1	0	0	0	1
2	0	1	0	1	0
…	…	…	…	…	…
N	0	0	1	1	0

Discretization

Figure 1. Binarization.

2.2. Rule creation

Association rules were created, represented as $X \rightarrow Y$, where X is the cause subset for machine failure, and Y is the subset for machine type. The itemsets of association rules become the predictive model for machine failure. To create the association rules, frequent itemsets should be extracted from the binarized dataset. We used the Lattice model, as shown in Figure 2, to list all possible itemsets (i.e., candidates), and the itemset of each machine in the binarized dataset was compared with every candidate to identify frequent itemsets. However, this procedure may lead to significant computational burden on the system, since the number of comparisons increases exponentially as the number of items increase. Therefore, we applied the Apriori algorithm to reduce the number of candidates compared. The Apriori algorithm follows two principles: i) If an itemset occurs frequently, then all of its subsets must also be frequent, and ii) If an itemset occurs infrequently, then all of its subsets must also be infrequent. This reduces the number of comparisons by pruning subsets of infrequent itemsets, as shown in Figure 3. Frequent items can be identified depending on the value of support, confidence, and lift;

$$s(X \rightarrow Y) = \frac{n(X \cup Y)}{N}, \tag{1}$$

$$c(X \rightarrow Y) = \frac{n(X \cup Y)}{n(X)}, \tag{2}$$

and

$$Lift(X,Y) = \frac{c(X \rightarrow Y)}{s(Y)}, \tag{3}$$

respectively; where N is the number of machines, $n(X \cup Y)$ is the number of itemsets including X and Y, $n(X)$ is the number of itemsets including X, and $s(Y)$ is the support for Y.

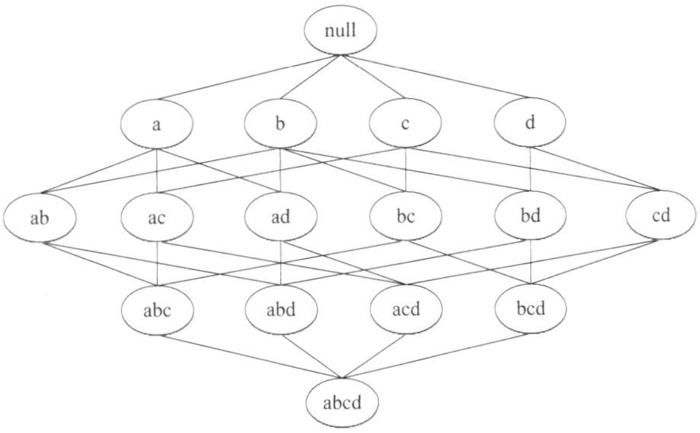

Figure 2. Lattice model.

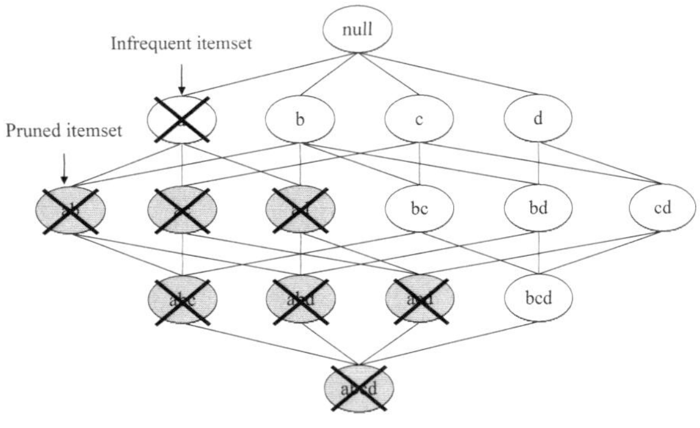

Figure 3. Itemset pruning.

2.3. Visualization

Finally, the created association rules were visualized in various ways, e.g. graphs, tables, matrixes, etc., to help the end user understand the rule patterns. Visualization included sorting and filtering to simplify finding specific rules.

IOP Conf. Series: Journal of Physics: Conf. Series **892** (2017) 012008 doi:10.1088/1742-6596/892/1/012008

3. Implementation

To evaluate the feasibility of the proposed machine failure prediction model, we conducted an experimental implementation using R Studio version 3.3.2. In the following subsections, we describe the dataset and discuss the results of the implementation, in detail.

3.1. Dataset

We used log data collected from 1,000 machines, consisting of the failure cause and failure type items for each machine. The cause items were CPU usage ratio, memory usage ratio, machine temperature, and residual battery; and the type items were long latency, shutdown, and freezing.
CPU and memory usage ratios are continuous items, and are commonly divided into three categories: low, middle, and high level. However, the category ranges are different: CPU usage ratio categories were [0, 30], (30, 50], and (50, 100]; whereas memory usage ratio categories were [0, 50], (50, 80], (80, 100], respectively. Machine temperature was divided into two categories: low and high, being < 60°C and ≥ 60°C, respectively. Other categorical items were represented as 1 or 0, corresponding to Yes or No, respectively.

3.2. Implementation results

Table 1 summarizes the implementation when support, confidence, and lift for frequent items were greater than 0.07, 0.5, and 2, respectively. In the table, the left-hand-side (LHS) lists the cause itemsets for machine failure, and the right-hand-side (RHS) lists the itemsets for machine failure type. Eleven association rules were identified. Only low battery affects machine shutdown, and machine freezing occurred when the machine temperature and CPU usage ratio become high. Long latency was affected by multiple items, including CPU usage ratio, memory usage ratio, and machine temperature, and others, so it is difficult to identify the cause of a specific long latency. Table 2 shows the association rules when support, confidence, and lift for frequent items were greater than 0.09, 0.5, and 2, respectively. The number of association rules is less than the previous case (Table 1). There are only 3 association rules for long latency, and we can identify that CPU and memory usage ratio have the largest impacts on long latency.

Table 1. Association rules (Support: 0.07, Confidence: 0.5, Lift: 2).

LHS	RHS	Support	Confidence	Lift
{Low.battery=O}	{Shut.down=Y}	0.099	1	10.101
{CPU_cd=CPU_high, Temperature_cd=Temperature_high}	{Freezing=Y}	0.099	0.831	8.403
{CPU_cd=CPU_mid, Memory_cd=Memory_mid}	{Long.latency=Y}	0.088	0.838	3.862
{CPU_cd=CPU_mid, Temperature_cd=Temperature_low}	{Long.latency=Y}	0.087	0.527	2.429
{Low.battery=X, CPU_cd=CPU_mid}	{Long.latency=Y}	0.139	0.522	2.408
{Memory_cd=Memory_mid, Temperature_cd=Temperature_low}	{Long.latency=Y}	0.109	0.521	2.403
{Low.battery=X, CPU_cd=CPU_high, Temperature_cd=Temperature_high}	{Freezing=Y}	0.099	0.942	9.523
{Low.battery=X, CPU_cd=CPU_high, Temperature_cd=Temperature_low}	{Long.latency=Y}	0.075	0.528	2.433
{Low.battery=X, CPU_cd=CPU_mid, Memory_cd=Memory_mid}	{Long.latency=Y}	0.088	0.907	4.180

IOP Conf. Series: Journal of Physics: Conf. Series **892** (2017) 012008 doi:10.1088/1742-6596/892/1/012008

{Low.battery=X, CPU_cd=CPU_mid, Temperature_cd=Temperature_low}	{Long.latency=Y}	0.087	0.583	2.690
{Low.battery=X, Memory_cd=Memory_mid, Temperature_cd=Temperature_low}	{Long.latency=Y}	0.109	0.567	2.616

Table 2. Association rules (Support: 0.09, Confidence: 0.5, Lift: 2).

LHS	RHS	Support	Confidence	Lift
{Low.battery=O}	{Shut.down=Y}	0.099	1	10.101
{CPU_cd=CPU_high, Temperature_cd=Temperature_high}	{Freezing=Y}	0.099	0.831	8.403
{Low.battery=X, CPU_cd=CPU_mid}	{Long.latency=Y}	0.139	0.522	2.408
{Memory_cd=Memory_mid, Temperature_cd=Temperature_low}	{Long.latency=Y}	0.109	0.521	2.403
{Low.battery=X, CPU_cd=CPU_high, Temperature_cd=Temperature_high}	{Freezing=Y}	0.099	0.942	9.523
{Low.battery=X, Memory_cd=Memory_mid, Temperature_cd=Temperature_low}	{Long.latency=Y}	0.109	0.567	2.616

Figures 4–7 visualize the created association rules. Figures 4 and 5 show the graph for 11 and 6 rules (corresponding to Tables 1 and 2), respectively, where green circles are items, and pink circles represent association rules where the lift is indicated by the circle size. Rule relationships are represented by arrows and circle positions. Figures 8 and 9 show the grouped matrixes for the same association rules, respectively, where the item support is represented by the circle size, and lift by circle color.

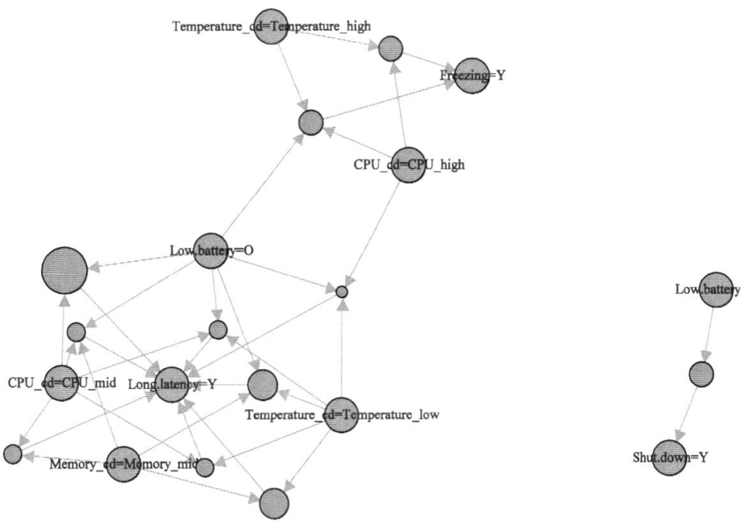

Figure 4. Graph for 11 rules.

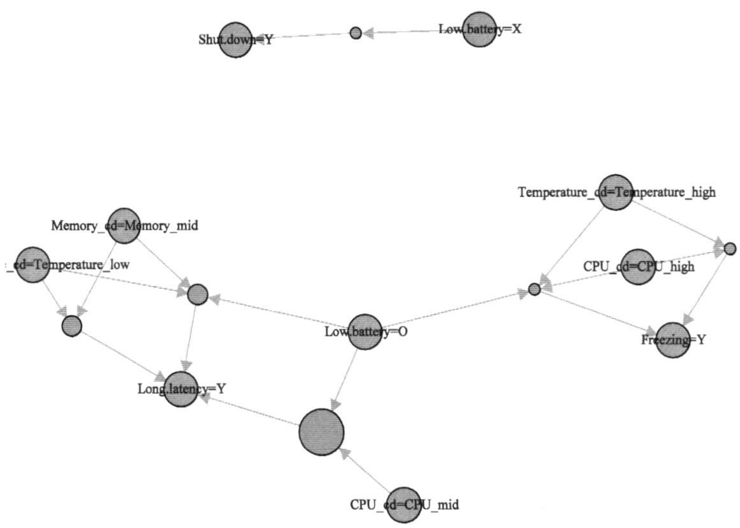

Figure 5. Graph for 6 rules.

Figure 6. Grouped matrix for 11 rules.

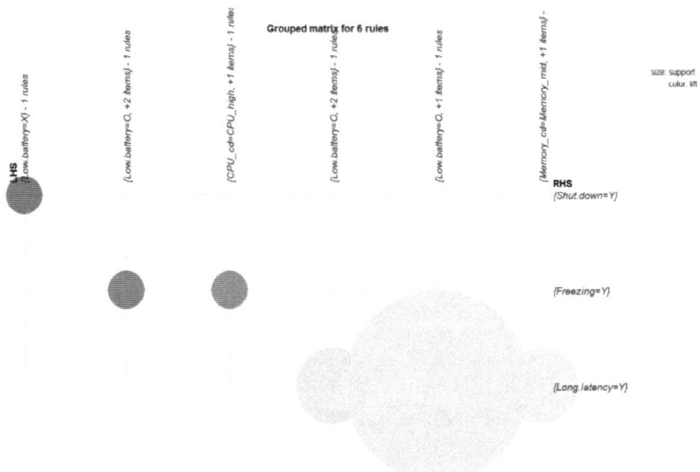

Figure 7. Grouped matrix for 6 rules.

4. Conclusions

We proposed an association rule-based predictive model for machine failure in industrial Internet of things (IIoT), which provides realistic association rules. Three major steps were considered in developing the model: 1) binarization, 2) rule creation, 3) visualization. Binarization transforms item values to a binary representation, and association rules were created using the Lattice model and Apriori algorithm. Visualization displays the identified rules in various ways. An experimental predictive model was implemented using R Studio version 3.3.2, and the identified association rules visualized using tables, graphs, and grouped matrixes. The implementation verified that the proposed predictive model for machine failure can provide realistic prediction results.

Acknowledgements

This work was supported by the Hallym Leading Research Group Support Program of 2017 (HRF-LGR-2017-0003).

References

[1] Hossain M S and Muhammad G 2016 *Comput. Netw.* **101** 192–202

[2] Posada J, Toro C, Barandiaran I, Oyarzun D, Stricker D, de Amicis R and Vallarino I 2015 *IEEE Comput. Graph. Appl. Mag.* **35** 26–40

[3] Kwon J -H, Lee H -H, Lim Y and Kim E -J 2016 *Appl. Sci.-basel* **6** 427:1–427:17

[4] Hashem I A T, Yaqoob I, Anuar N B, Mokhtar S, Gani A and Khan S U 2014 *Inform. Syst.* **47** 98–115

[5] Sahoo S K, Rodriguez P and Savinovic D 2015 *Proc. IEEE Int. Conf. on Electric Machines & Drives Conference (IEMDC)* pp 1398–1404

[6] https://www.rstudio.com/

[7] Zaki M J, Meira Jr W and Meira W, 2014 Itemset mining *Data mining and analysis: fundamental concepts and algorithms* (Cambridge: Cambridge University Press) chapter 8 pp 240–263

IOP Conf. Series: Journal of Physics: Conf. Series **892** (2017) 012009 doi:10.1088/1742-6596/892/1/012009

Replacing missing values using trustworthy data values from web data sources

M.Izham Jaya[1], Fatimah Sidi[1], Sharmila Mat Yusof [1,2], Lilly Suriani Affendey[1], Iskandar Ishak[1] and Marzanah A. Jabar[1]

[1]Faculty of Computer Science and Information Technology, Universiti Putra Malaysia, Serdang, Selangor, Malaysia.
[2]School of Computing, College of Arts and Sciences, Universiti Utara Malaysia, Sintok, Kedah, Malaysia.

fatimah@upm.edu.my

Abstract. In practice, collected data usually are incomplete and contains missing value. Existing approaches in managing missing values overlook the importance of trustworthy data values in replacing missing values. In view that trusted completed data is very important in data analysis, we proposed a framework of missing value replacement using trustworthy data values from web data sources. The proposed framework adopted ontology to map data values from web data sources to the incomplete dataset. As data from web is conflicting with each other, we proposed a trust score measurement based on data accuracy and data reliability. Trust score is then used to select trustworthy data values from web data sources for missing values replacement. We successfully implemented the proposed framework using financial dataset and presented the findings in this paper. From our experiment, we manage to show that replacing missing values with trustworthy data values is important especially in a case of conflicting data to solve missing values problem.

1. Introduction

Data is a very important substance in making analysis. However, collected data is usually incomplete and contains missing values. In practice, missing values can happen due to several reasons such as human errors, equipment malfunction, manual data entry process and incorrect measurement [1,2]. As missing values is a common problem in collected data, adopting suitable approach in managing missing values is crucial to minimize bias in the resulted analysis [3].

In data quality research, the occurrence or absence of missing values in which data records has the undesirable null values determined the completeness of data [4–6]. Due to this definition, we can consider that data is complete when all necessary values pertaining to the data exist and contained no missing values [7–9]. The emergence of research in managing missing values problem proposed various approaches to impute missing values. Most of the approach introduced a new value to replace missing value and further, make the data complete. A common example is to replace missing values with the observed mean. More robust statistical methods such as regression analysis have also been used to impute missing values efficiently.

Introduction of new values after missing values imputation using statistical methods brought several problems such as data duplication, data outliers and data distortion if it is not done by the statistical expert [10,11]. Consequently, the resulted analysis of the imputed data is deteriorate. The other way to find suitable values to replace missing values is to look for the data values that described about the same data in the other data sources such as web. However, as the missing values being replaced with data values from web, question on the level of trust that we have towards the data value retrieved from web need to be address. Obviously, replacing missing values with data values from other data source without knowing how much we can trust that data values risk the resulted analysis and worst, lead to a wrong decision.

Web contained large collection of potential data candidates to replace missing values. For example, in financial domain, web data from Yahoo! Financial and Google Finance can be used to replace missing values in financial statements dataset. However, heterogeneity in web data sources poses challenges to integrate and use web data to replace missing values. Furthermore, various schemas is used to define data in web. Thus, problems such as conceptual inaccuracies and terminological ambiguity is inevitable [12]. Ontology is used in [12] to solve the problems mentioned before.

In web, values of the same data from various web data sources can be different and conflicting with each other [13]. Replacing missing values with data value from web raised concern about the level of trust that we have towards the web data. Essentially, the selection of trustworthy web data to replace missing values derived a trusted completed dataset [6,14]. Moreover, by knowing the trust level of the selected data, users have more confident in the analysis result. The question of trust as mention before motivates us in doing this research.

Any replacement of missing values with data values from web required the calculation of trust score beforehand. Data candidate with highest trust score is then use to replace missing values. In this paper, we proposed a framework to manage missing values by replacing it with a trustworthy data values from web data sources. Therefore, we also introduced a trust measure to calculate trust score of each data candidates. The proposed trust measure is based on the available observed data in the dataset. In this paper, we use data candidate as a term that refer to the data values from web data sources and has potential to replace missing values.

This paper is structured as follows: in Section 2, we review related works in missing values mechanism, missing values imputation, ontology adoption in data quality, data accuracy and data reliability in measuring trust. In Section 3, we proposed a framework to manage missing values by replacing it with a trustworthy data values from web data sources. Then, we show illustrative example of the proposed framework in Section 4 and we conclude our findings in Section 5.

2. Related Works

2.1. Missing values mechanism

Missing values occurrence can be categorized into three categories, which are Missing Completely at Random (MCAR), Missing at Random (MAR) and Missing Not at Random (MNAR) [15]. Missing values are considered as MCAR when its' occurrence is independent to any other observed values or the data of interest itself. For example, if the interviewer only asked 60 percent of the questions randomly, then 40 percent of data resulted from the interview is missing. As the missing values occurred in random and not dependent to any observed value, 40 percent of missing values from the interview session can be consider as MCAR. Any imputation methods used to impute missing values with MCAR type will not lead to bias during the analysis [2]. However, the assumption of MCAR should be done with cautious as MCAR type is rarely occur in the dataset. For this reason, Little MCAR test [16] should be conducted in order to confirm the assumption.

On the contrary, missing values is considered as MAR if the occurrence of missing values is dependable to the other observed values but not to the missing value itself. Additionally, Acock [17] explained that missing values can be explain by other observed values in the case of MAR since the missing value is dependent to other observed values which are not missing. As an example, during the interview, students with age 15 to 17 years old tend not to answer the question of smoking status. This lead to the assumption that chances of missing values in smoking status of the student with age 15 to

17 years old is higher compared to the students with younger age. The resulted missing values from the interview can be consider as MAR since the missing value in smoking status is related to the observed variable, students' age. In MAR, missing values can be imputed by using the values of other observed values [3].

MNAR occurs when the possibility of missing values in the dataset is related to the unobserved values. Due to this, missing values in MNAR cannot be imputed by using other observed values [3]. Therefore, advance statistical knowledge is important in order to analyze data with MNAR [18]. In summary, it is important to make a clear distinction between MCAR, MAR and MNAR occurrence in the dataset as it will determined the best approach to deal with missing values.

2.2. Approaches in missing values imputation

Previous research in improving data completeness proposed various approaches to deal and to impute missing values. The most common approach is list-wise deletion by omitting instances with missing values from the analysis which then reduced size of instances. The list-wise deletion is simple. Yet, as the sample size decrease, the statistical power of the analysis will drop as well which then arise difficulties to detect small effect or relationship between variables [18]. It became worst in a case of MCAR data type, as it increase the standard errors and decrease the level of significance due to a smaller sample size [17]. Conversely, pair-wise deletion gives more statistical power without threaten the sample size. In pair-wise deletion, data with missing values is not included in the analysis, but the case will not be omit as it is used to analyze other observed data with non-missing values. Either list-wise or pair-wise deletion, the execution of the deletion approach is highly suggested only in MCAR and in some MAR cases where missing values only comprises of a small part of data [17, 19, 20].

Another approach to deal with missing values is to replace missing values with the value of observed mean [21]. In this approach, the mean value of the observed variables in dataset is computed and used to replace missing values. This approach is easy to use and more assertive compared to the deletion approach as it estimates the missing values based on the non-missing values without reducing the sample size and disturbing the mean variable of dataset [18]. However, using constant mean value to replace missing values in dataset distort the statistical distribution of data and will further risk the result of data analysis [22]. Besides, using the constant value of mean to replace missing values altered the randomness of data. Escaping the data randomness led to invalid statistical inference [18, 23].

In contrast to the abovementioned methods, Batista and Monard [24] proposed an approach based on the KNN algorithm to impute missing values. In order to find the nearest neighbor, several distance function such as Euclidean, Manhattan and Pearson is used. KNN suggested a set of k-nearest neighbor of the missing value which then being used to replace the missing value. Thus, the optimal value of k is important to ensure high imputation performance. Selection of the replacement techniques to replace missing values is depending on the type of data being analyzed. The most common techniques are mean estimation, median, mode and weighted average of the values [25]. As opposed to the mean imputation method, KNN can be used to predict missing values in both qualitative and quantitative data type [26]. However, finding the most similar instances in a large dataset is a time consuming process as the approach will search throughout the dataset [24].

Another approach to impute missing values is the MissForest method and it is found to be more robust compared to KNN [27,28] and mean imputation [28,29]. MissForest method imputes missing values based on the random forest algorithm which been trained on the non-missing values of the dataset [27]. Missing values imputation in MissForest is done in a repeated iteration which started by initial guess of the missing value, X by using mean imputation or other imputation methods. A random forest model is then created for each value of X to impute the missing value. The iteration is then stopped after the stopping criteria is met. Stopping criteria is met when a large differences between previous and the new imputed data matrix occur. The performances of MissForest is assessed by the Normalize Root Mean Squared Error (NRMSE) and Proportion of Falsely Classified Entries (PFC) for categorical variable which scaled from 0 to 1. Value of NRMSE and PFC that closest to 0 shows a good performance of MissForest. The ability of MissForest to handle any type of data with small tuning and produced better imputation result were the main advantage of this approach compared to others [28].

Further advancement in computing has resulted new approaches in improving data completeness and missing values including constraint based method [30] and replacing missing values with data from online data sources [12]. Prediction model for asset valuation has been used in [12] to evaluate the proposed framework. Various imputation approaches such as mean of available observation, ignore missing values and fuzzy concept is used to impute the dataset and later, being compared with the proposed framework in the evaluation. The result proved that replacing missing values with data from online data sources produced higher accuracy in the prediction model compared to other imputation approaches.

2.3. Ontology adoption in data quality

Data from various web data sources were usually found conflicting with each other [13]. The most common example is the difference of terms used to describe about the same data. Moreover, conflicting data from various web data sources can be resulted from the lack of precision in concept definitions. Du and Zhou [12] described both problems as terminological ambiguity and conceptual inaccuracy. Ontology adoption is important to solve both problems and to provide inference before data from web data sources is use to replace missing values.

Ontology has been used in previous research to solve data quality and data completeness problem. Several reasons strengthen the needs to adopt ontology in data quality such as the ability to embed domain knowledge and user's data quality requirement [31]. Data quality requirements explained how the assessment will be carried out including details about data quality metrics and the assessment method in used. Furthermore, the adoption of ontology in data quality management reduces extensive involvement of domain expert and data users during data quality assessment and improvement process. Additionally, previous researchers adopted ontology in data quality assessment because its ability to infer and to represent data from heterogeneous data source or data schema [12, 14], [31–35]. The adoption of ontology also allowed data quality assessment of large data to be conducted without expert involvement [12, 30], [32–34].

Ontology is an explicit specifications of concepts and it showed the relationship among concepts and their attributes in a specific domain [36]. In ontology, class is used to explain concepts in the interested domain. Each classes contained instances that specifically described the object. For example, a class of cars represents all cars and specific car model is the instances. Ontology can be built manually, semi-automated or fully automated. However, the process involved in building the ontology itself is complicated and time consuming. Thus, existing ontology that described interested domain can be used to safe time. In some cases, small modification or mapping of existing ontology is needed to suit the needs.

2.4. Measuring trust: data accuracy and data reliability

Dimensions of data accuracy and data reliability can be categorized into two perspective of trust which are: trust of data value and trust of data source. Trust of data value explained the dimensions that are important in measuring trust of the data value itself. The later described the measurement of trust in the data source that the value being retrieved. In this paper, focus is given on data accuracy and data reliability dimensions to measure trust score as our literature review suggested that data accuracy and data reliability is important in measuring trust and has been discussed in most research articles that emphasis trust [6, 32], [37–40].

Data accuracy is defined as the closeness of data value to the real world or the value of reference [41, 42]. However, in most cases it is difficult for us to know the value of data in real world. Thus, research in data quality categorized data accuracy into two categories which are: syntactic and semantic accuracy [6, 40]. In syntactic accuracy, the real value in real world data is not important in measuring accuracy. However, the semantic accuracy measures the closeness of data value to its real value. We can relate syntactic accuracy to the definition of accuracy by Redman [43] that data accuracy measure the proximity of data value from the value of reference that is considered correct. Data accuracy can be measure using distance function [6] where the closest distance between data value, v and value of reference v' reflected high data accuracy. In determining the accuracy of data source, the ratio of number of values that accurate and the number of total data values is used [40].

Data accuracy is related to the reliability of data source. If the data value is accurate, data accuracy is equal to 1 and its' data source is considered as highly reliable [39]. Thus, data source with lowest data accuracy rate, has low data reliability score. From data source perspective, data reliability is defined as the ability of data source to provide rightful data [6]. Instead, data reliability from data value perspective can be define as a measure of data ability to be trusted and conveys right information [6, 44]. Obviously, both definitions relate data reliability with trust and rightful data.

Reliability can be measure using several approaches such as Cronbach alpha and composite reliability score [37], trust measure [38] and truth discovery [39]. In discovering truth and determining data reliability, data which claimed by majority of data sources is considered as truthful data [39]. However, the proposed approach does not distinguish the reliability of data sources that claimed the same data value. It is important to expose the reliability of data sources as the majority claim can be wrong especially in the event that data being shared among data sources. Li et. al [39] measure source reliability and the confidence interval of the invariance to discover truth.

3. Framework of Missing Values Replacement Using Trustworthy Data Values from Web Data Sources

In this paper, we are interested in finding trustworthy data values from web data sources to replace missing values in dataset. We propose a framework to manage missing values by replacing it with a trustworthy data values from web data sources. Hence, we illustrated the proposed framework in figure 1.

The web contained a lot of data related to the domain of interest. For instance, in financial domain, Yahoo! Finance and Google Financial stored companies' financial data such as the income statement, balance sheet and share prices. However, making used of these data to replace missing values required the adoption of ontology to map the data from web data sources to the original dataset.

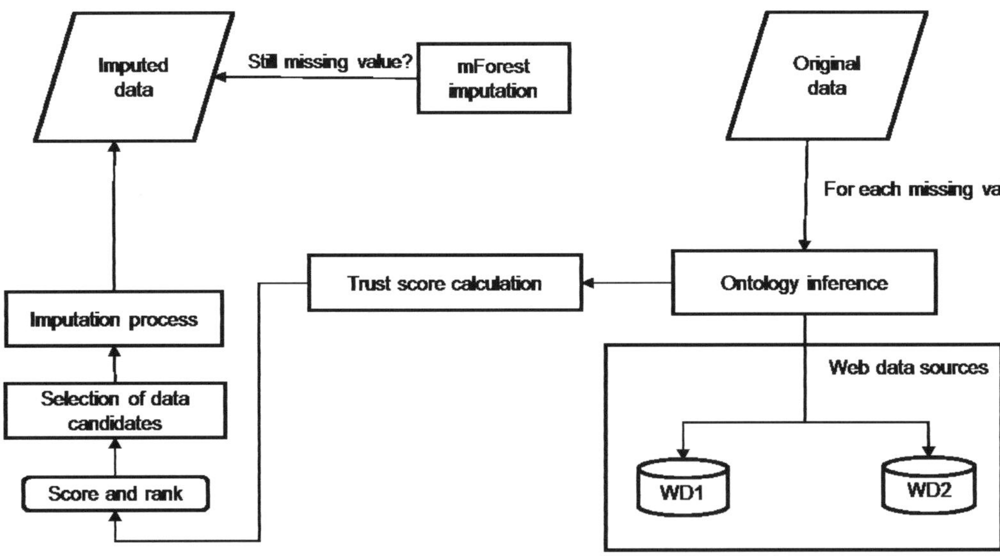

Figure 1. A framework to replace missing values with a trustworthy data values from web data sources.

For example, if a value of sales from company A is found missing, the ontology is then used to map related data from the web to the original dataset. In some cases, the term sales is not used to

IOP Conf. Series: Journal of Physics: Conf. Series **892** (2017) 012009 doi:10.1088/1742-6596/892/1/012009

represent data about sales. Instead, revenue is used to represent sales data. Conflicting terminology in web data sources is the example of terminological ambiguity [12]. The ontology is also use to infer data from web data sources and suggested suitable data candidates to replace missing values. As shown in figure 1, for each missing values found in the original data, the ontology is used to infer data from web data source 1 (WD1) and web data source 2 (WD2). Trust score is then calculated for data candidates from each web data sources. Data candidate with highest trust score is then selected to replace the missing values. If there is no values being retrieved from WD1 and WD2, then the missing values will be impute using MissForest [27]. The details of how to calculate the trust score is explained in the next section.

3.1. Trust score calculation

Data values retrieved from web data sources is usually conflicting with each other [13]. Importantly, only one data value is required to replace the missing value in the dataset at one time. It is impossible for us to know what are the real value that missing. Therefore, in order to select which data values should be used to replace missing value, we calculated the trust score. We define the following measure to calculate the trust score.

Assume that we have a financial dataset of twenty rows and four column. Each rows described financial data of the company (entity) and each column is the related variables. The dataset has five missing values. We retrieved data candidate from web data source (i) to replace missing value (x,j) in the dataset. The calculation of trust score for data candidate from web data source (i) is defined as:

$$100 * Acc_{wd(i)} * Rel_{wd(i)} * D_{variable}^{Relwd(i)} \tag{1}$$

Where,
$Acc_{wd(i)}$ = Accuracy score of data value from web data source (i)
$Rel_{wd(i)}$ = Reliability score of data value from web data source (i)
$D_{variable}$ = Average distance between 3 available variable in web data source (i) and the corresponding variables in original data

The calculation of trust score relied on the accuracy and reliability score of data candidate from web data source (i). As trust score measure the trustworthy of data candidates from web data source (i), high accuracy and reliability score of data candidate resulted high trust score. $D_{variable}$ calculated the average distance between data values of three available variables in web data source (i) and its corresponding variables in original dataset. Importantly, these three available data values belongs to the same entity that have the missing value. Thus, high discrepancies between values resulted high $D_{variable}$ values. We defined $D_{variable}$ measure as:

$$\left(\frac{\sum \left(A(i) - A(ori)\right)^2 + \left(B(i) - B(ori)\right)^2 + \left(C(i) - C(ori)\right)^2}{3} \right)^{1/2} \tag{2}$$

Where,
A(i), B(i), C(i) = Variable A, B, C from Source (j) which belongs to the same entity that have the missing value (x,j)
A(ori), B(ori), C(ori) = Variable A, B, C from original data which belongs to the same entity that have the missing value (x,j)

We explain the calculation of accuracy score, reliability score in the following subsection.

3.1.1. Calculation of accuracy score. The calculation of accuracy score is based on the similarity between values from three nearest variables in the original dataset and the values of corresponding variables retrieved from web data source (i). The value A from web data source (i) is considered as accurate if it similar with the corresponding value in the original data. Thus,

$$A_{\text{web data source (i)}} - A_{\text{original}} = 0 \tag{3}$$

The following measure defined how we calculate the accuracy score.

Assume that we want to replace a missing value Original(x,j) in our original dataset. The dataset comprises of twenty companies (entity) and four columns (variables). The nearest available variables to the missing value in the original dataset is A(x,k), B(x,l) and C(x,m). We retrieved data values of variables A, B and C from web data source (i) to measure the accuracy of its data values. The calculation of accuracy score is defined as:

$$\left(\frac{|A(i)|}{n}+\frac{|B(i)|}{n}+\frac{|C(i)|}{n}\right)/3 \tag{4}$$

Where,
|A(i)|, |B(i)|, |C(i)| = Number of entity from Variable A, B, C of Source (j) which the similarity is 0
n = Number of entity in dataset

3.1.2. Calculation of reliability score. A particular web data source is considered unreliable if it has too many data values that is different from the available observed data in the original dataset. In such cases, the variance of the error distribution in the particular web data source is big [39]. In this paper, a data that is retrieved from a web data source that has lower reliability score is considered as less reliable compared to the data from a web data source with higher reliability score.

Assume that we want to replace a missing value Original(x,j) in our original dataset. The dataset comprises of twenty companies (entity) and four columns (variables). The nearest available variables to the missing value in the original dataset is A(x,k), B(x,l) and C(x,m). We retrieved data values of variables A, B and C from web data source (i) to measure the reliability of its data values. The calculation of reliability score is defined as:

$$1/\left(\frac{x^2}{\sum\left(A(i)-A(ori)\right)^2+\left(B(i)-B(ori)\right)^2+\left(C(i)-C(ori)\right)^2}\right)_{n\in Ni} \tag{5}$$

Where,
x2 = Chi-square (0.025, df=20)
n = entity n
Ni = Set of entity provided in web data source (i)
A(i), B(i), C(i) = Variable A, B, C from web data source (i)
A(ori), B(ori), C(ori) = Variable A, B, C from original dataset

4. Illustrative Example

We present the example of our proposed framework using financial dataset obtained from Standard & Poor's Compustat North America dataset. The data were extracted for the first quarter 2009. The dataset comprises of twenty companies randomly selected with four variables which are: selling, sales, net income and gross profit. Data values in each variables explained total in million dollars for the respective company in the first quarter of 2009. Among the selected variables, sales, net income and gross profit has no missing values. Whereas, five companies has missing values in the selling variable.

We build the ontology in Protégé 4.3 which used to map data values from the web data sources to the data values in the original dataset. All the measurement discussed before is coded in R environment. In this example, two financial website is used to retrieve candidates to replace missing data. It is noted from table 1 that the retrieved data from both financial website exhibited conflicting values in selling variable. Thus, it justified the needs to measure the trust score before any replacement of missing values being made.

Table 1. Conflicting values in selling variables.

Company	Original dataset	Financial website 1	Financial website 2
ABT	Missing	2230	2225
NBR	Missing	69.28	69.30
ADSK	Missing	283.4	47.1
BHI	Missing	283	283
BAX	Missing	788	788

The resulted study showed that all data values retrieved from financial website 1 is given highest trust score compared to data values from financial website 2. It is shown in table 2 that the accuracy score and $D_{variable}$ in data values from both financial websites is not much different, but as the reliability score between both financial website has a wide gap, we can see a large different between their trust score.

Table 2. Resulted trust score for each missing values.

Company	Average Distance		Reliability		Accuracy		Trust Score	
	FW 2	FW 1	FW 2	FW 1	FW 2	FW 1	FW 2	FW 1
ABT	0.49	0.50	0.25	0.74	0.30	0.27	6.32	12.05
NBR	0.50	0.49	0.25	0.74	0.30	0.27	6.23	12.03
ADSK	0.50	0.50	0.25	0.74	0.30	0.27	6.33	12.03
BHI	0.39	0.60	0.25	0.74	0.30	0.27	5.95	13.94
BAX	0.49	0.50	0.25	0.74	0.30	0.27	6.32	12.09

In comparisons to the original data, financial website 1 has a total of sixteen similar data values and total errors of 8735.23 in sales, net income and gross profit. In order to calculate the total errors, we measured the discrepancies between each data values of sales, net income and gross profit from web data source and the original data. On the other hand, a total of eighteen data values from financial website 2 has been identified as similar to the original data, but higher total errors of 11279.33 is recorded. As shown in table 2, the accuracy of data values retrieved from financial website 2 is higher than financial website 1. The high number of similar data values retrieved from financial website 2 compared to financial website 1 explained the findings. However, the reliability score of data values from financial website 2 is much lower as higher total errors is recorded in data values retrieved from financial website 2.

We ranked the trust score of each data values retrieved from financial website 1 and financial website 2 in table 3. Based on the ranking, we then selected data values with highest trust score to replace missing values in the original dataset. In this example, all the missing values have been replace with data values retrieved from financial website 1.

Table 3. Ranking of data values from web data source according to trust score.

	Ranking		Selected data value
	First	Second	
Missing value 1	financial website 1	financial website 2	2230
Missing value 2	financial website 1	financial website 2	69.28
Missing value 3	financial website 1	financial website 2	283.4
Missing value 4	financial website 1	financial website 2	283
Missing value 5	financial website 1	financial website 2	788

5. Conclusion

In this paper, we propose a framework to manage missing values by replacing it with a trustworthy data values from web data sources. The framework address the problem of replacing missing values with conflicting values from web data sources by providing trust score measurement before any replacement being made. Furthermore, the trust score provided in this framework facilitate user to gain confidence in using the completed dataset for data analysis.

The proposed framework is our initial work towards trusted missing values management approach. Way forward, evaluation of the proposed approach should be carry out by analyzing the usability of the proposed approach in MCAR, MAR and MNAR mechanism. In view that reliability and accuracy played important role in determining the trust score, further study can be done to analyze any correlation between reliability, accuracy and trust score.

References

[1] R. Deb and A. W.-C. Liew Missing value imputation for the analysis of incomplete traffic accident data 2016 *Information Sciences* **339** pp 274–289
[2] C.-F. Tsai and F.-Y. Chang Combining instance selection for better missing value imputation 2016 *Journal of Systems and Software* **122**, pp 63–71
[3] G. Baio and B. Leurent An Introduction to Handling Missing Data in Health Economic Evaluations 2016 *Care at the End of Life* **39** Cham: Springer International Publishing pp 73–85
[4] Y. Liu, J. Li, and Z. Zou Determining the Real Data Completeness of a Relational Dataset 2016 *Journal of Computer Science and Technology* **31** no. 4 pp 720–740
[5] A. Wechsler and A. Even Assessing Accuracy Degradation Over Time With a Markov-Chain Model 2012 *Proceedings of the 17th International Conference on Information Quality (ICIQ-12)* pp 99–110
[6] C. Batini and M. Scannapieca 2006 *Data Quality* (Springer Berlin Heidelberg)
[7] V. Jayawardene, S. Sadiq, and M. Indulska An Analysis of Data Quality Dimensions 2013 *ITEE Technical Report No. 2013-01* **1** pp 1–32
[8] M. Bovee, R. P. Srivastava, and B. Mak A conceptual framework and belief-function approach to assessing overall information quality 2003 *International Journal of Intelligent Systems* **18** no. 1 pp 51–74
[9] B. K. Kahn, D. M. Strong, and R. Y. Wang Information quality benchmarks: product and service performance 2002 *Communications of the ACM* **45** no. 4 pp 184–192

[10] S. Sadiq 2013 *Handbook of Data Quality* (Berlin, Heidelberg: Springer Berlin Heidelberg)
[11] T. Dasu and J. M. Loh Statistical Distortion: Consequences of Data Cleaning 2012 *Proceeding of the VLDB Endowment* **5** no. 11 pp 1674–1683
[12] J. Du and L. Zhou Improving financial data quality using ontologies 2012 *Decision Support Systems* **54** no. 1 pp 76–86
[13] X. Yin, J. Han, and P. S. Yu Truth discovery with multiple conflicting information providers on the Web 2008 *IEEE Transactions on Knowledge and Data Engineering* **20** no. 6 pp 796–808
[14] X. Chu, J. Morcos, I. F. Ilyas, M. Ouzzani, P. Papotti, N. Tang, and Y. Ye KATARA: A Data Cleaning System Powered by Knowledge Bases and Crowdsourcing 2015 *Proceedings of the 2015 ACM SIGMOD International Conference on Management of Data - SIGMOD '15* pp 1247–1261
[15] R. J. A. Little and D. B. Rubin 1987 *Statistical Analysis with Missing Data* **2** 1987
[16] R. J. A. Little A Test of Missing Completely at Random for Multivariate Data with Missing Values 1988 *Journal of the American Statistical Association* **83** p.p 1198
[17] A. C. Acock Working With Missing Values 2005 *Journal of Marriage and Family* **67** no. 4 pp 1012–1028
[18] B. E. Cox, K. McIntosh, R. D. Reason, and P. T. Terenzini Working with Missing Data in Higher Education Research: A Primer and Real-World Example 2014 *The Review of Higher Education* **37** no. 3 pp 377–402
[19] J. L. Schafer and J. W. Graham Missing data: Our view of the state of the art 2002 *Psychological Methods* **7** no. 2 pp 147–177
[20] C. Paul, W. M. Mason, D. McCaffrey, and S. A. Fox A cautionary case study of approaches to the treatment of missing data 2008 *Statistical Methods and Applications* **17** no. 3 pp 351–372
[21] D. Michie, D. J. Spiegelhalter, C. C. Taylor, E. D. Michie, D. J. Spiegelhalter, and C. C. Taylor Machine Learning, Neural and Statistical Classification 1994 *Ellis Horwood series in artificial intelligence* **37** no. 4 p.p, 289
[22] J. Cao, X. Diao, N. Zhang, and T. Wang An Approach Using Hidden Markov Model for Estimating and Replacing Missing Categorical Data 2011 *Proceedings of the 16th International Conference on Information Quality (ICIQ-11)* no. 427 pp 427–434
[23] P. Royston Multiple imputation of missing values 2004 *The Stata journal* **4** no. 3 pp 227–241
[24] G. E. A. P. A. Batista and M. C. Monard A study of k-nearest neighbour as an imputation method 2002 *Frontiers in Artificial Intelligence and Applications* **87** pp 251–260
[25] L. Torgo, *Data Mining with R* 2010 **20105341** (Chapman and Hall/CRC)
[26] E. Acuña and C. Rodriguez The Treatment of Missing Values and its Effect on Classifier Accuracy 2004 *Classification, Clustering, and Data Mining Applications* Berlin, Heidelberg: Springer pp. 639–647
[27] D. J. Stekhoven and P. Bühlmann Missforest-Non-parametric missing value imputation for mixed-type data 2012 *Bioinformatics* **28** pp 112–118
[28] A. K. Waljee, A. Mukherjee, A. G. Singal, Y. Zhang, J. Warren, U. Balis, J. Marrero, J. Zhu, and P. D. Higgins Comparison of imputation methods for missing laboratory data in medicine 2013 *BMJ Open* **3** no. 8
[29] M. Misztal Some remarks on the data imputation 'MissForest' Method 2013 pp 169–179
[30] O. Curé Improving the Data Quality of Drug Databases using Conditional Dependencies and Ontologies 2012 *Journal of Data and Information Quality* **4** no. 1 pp 1–21
[31] N. Martin, A. Poulovassilis, and J. Wang A Methodology and Architecture Embedding Quality Assessment in Data Integration 2014 *Journal of Data and Information Quality* **4** no. 4 pp 1–40
[32] S. Destercke, P. Buche, and B. Charnomordic Evaluating Data Reliability: An Evidential Answer with Application to a Web-Enabled Data Warehouse 2013 *IEEE Transactions on Knowledge and Data Engineering* **25** no. 1 pp 92–105
[33] S. Geisler, S. Weber, and C. Quix Ontology-Based Data Quality Framework for Data Stream Applications 2011 *Proceedings of the 16th International Conference on Information Quality*

(ICIQ-11) pp 145–159

[34] L. Gao, M. Bruenig, and J. Hunter Semantic-based Detection of Segment Outliers and Unusual Events for Wireless Sensor Networks 2013 *Proceedings of the 18th International Conference on Information Quality (ICIQ-13)* pp 102–119

[35] C. Baillie, P. Edwards, and E. Pignotti QUAL: A Provenance-Aware Quality Model 2015 *Journal of Data and Information Quality* **5** no. 3 pp 1–22

[36] T. R. Gruber A translation approach to portable ontology specifications 1993 *Knowledge Acquisition* **5** no. 2 pp 199–220

[37] Y. Lu and D. Yang Information exchange in virtual communities under extreme disaster conditions 2011 *Decision Support Systems* **50** no. 2 pp. 529–538

[38] R. M. Rodríguez, D. Ruan, J. Liu, A. Calzada, and L. Martínez Imputing Missing Values in Nuclear Safeguards Evaluation by a 2-Tuple Computational Model 2010 *Artificial Intelligence and Soft Computing* pp 202–209

[39] Q. Li, Y. Li, J. Gao, L. Su, B. Zhao, M. Demirbas, W. Fan, and J. Han A confidence-aware approach for truth discovery on long-tail data 2014 *Proceedings of the VLDB Endowment* **8** no. 4 pp 425–436

[40] C. Batini, C. Cappiello, C. Francalanci, and A. Maurino Methodologies for data quality assessment and improvement 2009 *ACM Computing Surveys* **41** no. 3 pp 1–52

[41] R. Wang and D. Strong Beyond accuracy: What data quality means to data consumers 1996 *Journal of management information systems* **12** no. 4 pp 5–33

[42] D. P. Ballou and H. L. Pazer Modeling Data and Process Quality in Multi-Input, Multi-Output Information Systems 1985 *Management Science* **31** pp 150–163

[43] T. C. Redman 1996 *Data quality for the information age* (Boston: Artech House)

[44] Y. Wand and R. Y. Wang Anchoring data quality dimensions in ontological foundations 1996 *Communications of the ACM* **39** no. 11 pp 86–95

Selection Input Output by Restriction Using DEA Models Based on a Fuzzy Delphi Approach and Expert Information

Roslah Arsad[1, 3], Mohammad Nasir Abdullah[1], Suriana Alias[2] and Zaidi Isa[3]

[1]Faculty of Computer & Mathematical Sciences, Universiti Teknologi MARA, Perak Branch, Tapah Campus, 35400 Tapah Road, Perak Malaysia
[2]Faculty of Computer & Mathematical Sciences, Universiti Teknologi MARA, Kelantan Branch,18500 Bukit Ilmu, Machang, Kelantan, Malaysia
[3]School of Mathematical Sciences, Universiti Kebangsaan Malaysia, 43600 Bangi, Selangor, Malaysia

rosla280@perak.uitm.edu.my, nasir916@perak.uitm.edu.my, suria588@kelantan.uitm.edu.my, pghrisiko@ukm.edu.my

Abstract. Stock evaluation has always been an interesting problem for investors. In this paper, a comparison regarding the efficiency stocks of listed companies in Bursa Malaysia were made through the application of estimation method of Data Envelopment Analysis (DEA). One of the interesting research subjects in DEA is the selection of appropriate input and output parameter. In this study, DEA was used to measure efficiency of stocks of listed companies in Bursa Malaysia in terms of the financial ratio to evaluate performance of stocks. Based on previous studies and Fuzzy Delphi Method (FDM), the most important financial ratio was selected. The results indicated that return on equity, return on assets, net profit margin, operating profit margin, earnings per share, price to earnings and debt to equity were the most important ratios. Using expert information, all the parameter were clarified as inputs and outputs. The main objectives were to identify most critical financial ratio, clarify them based on expert information and compute the relative efficiency scores of stocks as well as rank them in the construction industry and material completely. The methods of analysis using Alirezaee and Afsharian's model were employed in this study, where the originality of Charnes, Cooper and Rhodes (CCR) with the assumption of Constant Return to Scale (CSR) still holds. This method of ranking relative efficiency of decision making units (DMUs) was value-added by the Balance Index. The interested data was made for year 2015 and the population of the research includes accepted companies in stock markets in the construction industry and material (63 companies). According to the ranking, the proposed model can rank completely for 63 companies using selected financial ratio.

1. Introduction

Stock selection to a portfolio has always been a major task and the investors have faced problem since the availability of the number of securities with multiple input and output data makes it more complex to determine the stocks. Therefore, the decision to evaluate and rank the stock can be determined using Data Envelopment Analysis (DEA). A variety of DEA models were formulated to rank the efficiency of Decision Making Units (DMUs). DEA is a mathematical decision making technique to measure the relative efficiency DMUs with multiple inputs and outputs. This method was originated by Charnes et. al [1]. One of the drawbacks of DEA is the problem of lack of discrimination among efficient DMUs and hence, yielding a large number of DMUs as efficient ones. Therefore two new models for ranking efficient DMUs based on *Norm 1* and using means of inputs-outputs weights were developed [2]. Cross efficiency evaluation method has been widely used for ranking the performance of DMUs in Data Envelopment Analysis. However, there were still several flaws in the cross efficiency evaluation method, such as non-uniqueness of cross efficiency scores, that may limit its usefulness [3].

Many methods have been proposed to enhance the discrimination power of DEA, one of which is the super-efficiency method proposed by Andersen and Petersen [4] to break the tie of efficient DMUs that occurs under the CCR model proposed by Charnes et al.[1]. However, super efficiency models have three drawbacks: first, the inability to rank non-extreme efficient DMUs; second, evaluating DMUs according to the different weight; and finally, efficiency models were infeasible in some cases [5]. Hosseinzadeh Lotfi et al. [6] proposed one DEA ranking method based on applying aggregate units. Selecting the best ranking method or the way of combining different ranking methods for ranking DMUs is an important point in ranking DMUs in DEA. Hence, they proposed methodology based on the TOPSIS method for ranking DMUs [7]. A different theories and methods will produce different ranking. Therefore, in this paper, we introduced a complete ranking of DMUs using restriction in DEA where the input and output are selected based on Fuzzy Delphi approach and previous studies. The model used for ranking is restriction method in DEA, proposed by Alirezaee and Afsharian [8]. This model is more precise and logical than other methods because it using initial data of DEA models and does not destroy competition between DMUs. Furthermore, it is easily employed and logical [8]. This model was modified by Dong Guo, Jie Wu [9] and Jie Wu [10]. Some traditional approaches to select attractive stocks by means of evaluating the overall corporate performance of the firm's financial statement analysis are based on the computation of financial ratio. These ratios provide a dynamic analysis of the company's financial situation and evaluate with precision the advantage and disadvantage of the company. Roslah et al. [11] adopted the model proposed by Alirezaee and Afsharian for ranking efficiency unit stocks performance in Bursa Malaysia where the inputs and outputs were selected based on the previous study.

Financial ratios were also used for the purpose of predicting future performance. However, using all the available financial ratios in stock evaluation is impractical as it will make the evaluation process computationally and analytically complicated. There were many methods which have been implemented to identify the most important ratios. In order to select and identify the most important financial ratio, Fuzzy Delphi Method (FDM) on expert judgement was proposed because the method is an analytical technique. The advantage of using FDM is that it requires only a small number of samples and the derived results are objective and reasonable. Therefore, it can save time and the cost of collecting expert opinion. More importantly, it takes into account the fuzziness that cannot be avoided during the survey process [12]. With application of Alirezaee and Afsharian's model, the DEA method was used to find the efficiency score and ranking on the selected stocks listed companies in Bursa Malaysia. CCR output oriented proposed calculating the efficiency score in every stock. The efficiency score has a range score between 0 and 1. A stock is efficient when the score is 1 and inefficient when the score is less than 1. If the stocks have the same efficient score, a problem will occur when ranking the stocks. Thus, the purpose of this study is to select the appropriate inputs and outputs, rank the stocks using the original model by Alirezaee and Afsharian using the Balance Index.

In order to perform the evaluation, financial analysis based on the computation of financial ratios can be used as evaluation criteria. Since there are many ratios available, it is important to identify the most critical ratio to be considered in the evaluation. In this study, we choose FDM to select the most important financial ratios for stock evaluation by using restriction in DEA. In the literature, many studies have used DEA model to measure performance companies and industry activities in various sectors. A summary of the literature including the DEA studies conducted by several business companies shown in Table 1. Based on Table 1, it shows that price to earnings ratio is the most parameter used among researchers to evaluate companies' performances. Therefore, in this research, we use the price to earnings ratio as one of the critical financial ratio to evaluate performance of companies.

Table 1. DEA studies for various business companies.

Authors	Business sector	Input variable	Output variable	DEA Model
[13]	10 largest industries in India	Price to earnings ratio (PER), Sigma (3 years), Beta	Earnings per share (EPS), 1 year return, 3 year return, 6 year return	SBM
[14]	28 cement's companies in Tehran stock market	Ratio to price income, Sigma (5 years)	Average of 5 year return on investment (ROI), dividends of stock	CCR
[15]	185 largest market cap stocks	Price earnings ratio (PER), Beta, Sigma (5 years)	Earnings per share (EPS), 1 year return, 3 year return, 5 year return, 10 year return	CCR
[16]	22 quarters Brazilian stock market	Price earnings ratio (PER) Beta, Return volatility	Earnings per share (EPS), 1 year return 3 year return, 5 year return	CCR
[17]	20 companies in Bursa Malaysia	Current ratio, Debt ratio, Debt to equity ratio	Return on investment (ROI), Return on equity (ROE), Earning per share (EPS)	BCC

2. Fuzzy Delphi Method and Expert Information

Delphi method is an interactive method used to survey and collect the most reliable consensus opinion of a group of experts on subject. To determine the most important financial ratios for stock evaluation, the ratios have been firstly identified from review of the existing literature. Twenty financial ratios were chosen based on their popularity and relevance to the assessment of stock performance. The selected financial ratios were categorized into five groups perspective namely liquidity, profitability, leverage, asset turnover ratios and growth ratios.

From the previous study, FDM was adopted to identify the critical financial ratios for evaluation of the stock performance. This method creates a better way of criteria selection. Seven linguistic terms expressed by triangular fuzzy numbers were used to indicate the level of importance of each ratio [12]. The seven linguistic terms were extremely unimportant (EU), unimportant (U), quite unimportant (QU), fair (F), quite important (QI), important (I) and extremely important (EI). In addition, their study had distributed 14 questionnaires to experts who work at investment companies, brokerage companies and academicians to obtain their opinions about the importance of the criteria. Mazura et al. [12] have applied FDM and the process were explained as follows:

Step 1: **Collect opinions of expert group**: Each expert is asked through a questionnaire to determine the importance of the evaluation criteria.

Step 2: Set up triangular fuzzy numbers: Calculate the evaluation value of the triangular fuzzy number of each criterion given by the expert. The evaluation value of a criterion j of m criteria given by expert i of n experts is expressed as $\tilde{w}_{ij} = (a_{ij}, b_{ij}; c_{ij})$, $i = 1,2,...,n$ and $j = 1,2,...,m.$

Then the overall fuzzy value $\tilde{w}_{ij} = (a_j, b_{j,} c_j)$ where $a_j = \min_j \{a_{ij}\}$, $b_j = \frac{1}{n}\sum_{i=1}^{n} b_{ij}$, $c_j = \max_j \{c_{ij}\}$

Step 3: Defuzzification: Use the center of gravity method to defuzzify the fuzzy value $\tilde{w}_j$ of each criterion to define value of S_j where $S_j = \dfrac{a_j + b_j + c_j}{3}$, $j = 1,2,...,m.$

Step 4: Screen evaluation criteria: To achieve the goal of screening, a threshold value α must be defined subjectively according to the needs of the study. The principle screening is as follows:

If $S_j \geq \alpha$, then the criterion is accepted.

If $S_j < \alpha$, then the criterion is rejected.

The evaluation criteria with a threshold value below 0.6 were deleted and the evaluation criteria with threshold value above 0.6 were selected [12]. From the research, the result identified the most important criteria for stock evaluation which are return on equity, return on asset, and earnings price per share, operating profit margin, net profit margin and debt to equity ratio. Table 2 shows seven critical financial ratio and definition for stock evaluation based on previous study and Fuzzy Delphi Method (FDM).

Table 2. Financial ratio for stock evaluation.

Ratios	Description
Return on equity	Net income divided by shareholder's equity
Return on assets	Net income divided by total assets
Earnings per share	Net income minus preferred dividends divided by common shares outstanding
Operating profit margin	Operating income divided by net sales
Net profit margin	Net income divided by net sales
Debt to equity ratio	Long-term debt divided by shareholders equity
Price to earnings ratio	Stock price divided by net income per share

Based on expert information, selection of inputs and outputs also will be clarified. Expert information (EI) on inputs and outputs selection in an industry may have been formed over many years of experiences of studying inputting and outputting. Researchers often refer to expert opinions in defining these variables. Normally, profitability and growth perspective are typically considered as outputs because revenue or income generation is a major objective criterion for a firm. On the other hand, asset utilization, liquidity and leverage perspectives are considered as inputs because they were concerned with planning and operational strategies of a firm. In contrast, valuation perspective was concerned with how well the equity markets perceived success of a firm and thus it is not concerned with a firm's input strategy [18]. Therefore, based on definition from expert information, we clarify the inputs and outputs. Table 3 shows classification of selected financial ratios as inputs and outputs. It is important to mention that, most of the selected ratios were under the profitability category. The results of this study can be used to observe the achievement of the companies in terms of financial performance. The financial ratios that has been identified were used as inputs and outputs to compute the efficiency score of stocks using the restriction in DEA model.

Table 3. Classification of financial ratio.

Ratios	Perspective	Type
Return on equity (ROE)	Profitability	Output
Return on assets (ROA)	Profitability	Output
Earnings per share (EPS)	Profitability	Output
Operating profit margin (OPM)	Profitability	Output
Net profit margin (NPM)	Profitability	Output
Debt to equity ratio (D/E Ratio)	Liquidity	Input
Price to earnings ratio (PER)	Valuation	Input

3. Model of Complete Ranking

The model proposed by Alirezaee and Afsharian's is presented as follows [8]. Their model considered the n decision making units which consume varying amounts of m inputs in the productions outputs. Then, suppose x_{ij} denotes the amount consumed by the ith input and y_{rj} denotes the amount produced by the rth output measured for the jth decision making unit. In order to measure the efficiency unit, DEA does not require any price data.

Let us define:

u_r :weight assigned to output r (r =1,...., s)

v_i : weight assigned to input i (i=1,......,m)

This method does not require any price data. Therefore, it uses vectors of shadow prices of inputs and outputs. Consider u and v are shadow price vectors for output and input respectively using these prices, efficiency for DMU$_p$ is:

$$EFF_p = \frac{\sum_{r=1}^{s} u_r y_{rp}}{\sum_{i=1}^{m} v_i x_{ip}} \quad (1)$$

There were two restrictions where the shadow prices must be non-negative and the shadow prices have to be such that when aggregated using these prices, no DMUs input-output bundle results in efficiency greater than unity. This also ensures that $\mathrm{EFF_p} \leq 1$ for DMU$_p$.

These restrictions can be formulated as follows:

$$EFF_j = \frac{\sum_{r=1}^{s} u_r y_{rj}}{\sum_{i=1}^{m} v_i x_{ij}} \leq 1, \quad \forall j \quad (2)$$

$$u_r, v_i \geq 0, \quad \forall r \forall i$$

The second restriction to get a profit for j^{th} DMU is presented as follows:

$$\left[\sum_{r=1}^{r} u_r y_{rj} - \sum_{i=1}^{m} v_i x_{ij}\right] \leq 0 \quad j = 1,.......,n. \tag{3}$$

When the shadow price was derived from technology, the input profits of DMU were zero. This situation is called a balanced situation. Therefore, when the profit restriction by shadow price becomes zero, we say that the pth DMU is efficient. The profit for other DMUs is equal to or less than zero, so the current DMU has overcome the others in this profit competition [8]. Otherwise, the DMU is inefficient because its profit restriction was not zero, thus, it was considered a loss. Next, Alirezaee and Afsharian have used the profit restriction and sum to describe a new index in addition to the efficiency score for each DMU. This situation is called Balance Index.

Normally, most of the sum was equal to or less than zero. Based on Dong Gua and Jie Wu [9], the less the sum is, the more greatly the profit of the evaluated DMU differs from the profits of the DMUs, and the higher the evaluated DMU should be ranked. Therefore, we can conclude that if the efficiency score unit of DMU_A and DMU_B is the same, and if DMU_A obtains more negative quantity values in Balance Index than DMU_B, then DMU_A has a better rank than DMU_B.

4. Methodology

This study has analysed 63 stocks in the construction and material industry on Bursa Malaysia. Table 4 shows the variable under consideration for each specific financial ratio. The Bursa Malaysia data for analysis was taken from Bloomberg and Datastream. Both Bloomberg and Datastream terminals were available at the Tun Abdul Razak Library (UiTM) Campus Tapah, Perak, Malaysia. The data on year 2015 were used to rank the performance of the selected companies.

All analysis were to determine the efficiency score. The software used for these analyses was R-programming software employing Benchmarking package. The efficiency scores of stocks were computed whereby the 2 input and 5 output data were considered. The method by Alirezaee and Afsharian [8] was applied in which the constant return to scale (CRS) was used. Once the values were obtained, the ranking were done using Balance Index.

Table 4. Variable for input, output and weight

Attribute	Value symbol	Weight symbol
Operating profit margin (OPM)	y_1	u_1
Earnings per share (EPS)	y_2	u_2
Net profit margin (NPM)	y_3	u_3
Return on equity (ROE)	y_4	u_4
Return on assets (ROA)	y_5	u_5
Debt to equity ratio (D/E Ratio)	x_1	v_1
Price to earnings ratio (PER)	x_2	v_2

IOP Conf. Series: Journal of Physics: Conf. Series **892** (2017) 012010 doi:10.1088/1742-6596/892/1/012010

The linear programming model for measuring the efficiency score the pth stock from 63 selected stocks were:

$$Max\ EFF_P = \sum_{r=1}^{5} u_r y_{rp} \tag{4}$$

Subject to the following constraints:

$$\sum_{r=1}^{5} u_r y_{rj} - \sum_{i=1}^{2} v_i x_{ij} \leq 0, \quad \text{for each stock } j \tag{5}$$

$$\sum_{i=1}^{2} v_i x_{ip} = 1, \quad \text{for each stock } j \tag{6}$$

$$u_r, v_i \geq 0, \quad \forall r \forall i \quad \text{for each stock } j$$

where u_r and v_i are the weights assigned to output and input respectively. Ranking the entire list of stocks completely, would be impossible if it were only based on the efficiency score. Therefore, we considered $\sum_{r=1}^{5} u_r y_{rj}$ as the total revenue and $\sum_{i=1}^{2} v_i x_{ij}$ as the total cost for 63 stocks which were calculated as an optimization problem.

$$\left[\sum_{r=1}^{5} u_r y_{rj} - \sum_{i=1}^{2} v_i x_{ij}\right] \leq 0 \quad j = 1,2,.......,63. \tag{7}$$

The Balance Index was computed for year 2015 as follows:

$$\text{Balance Index}_{2015} = [676.3u_1 + 416.8u_2 + 8.184u_3 + 525.4u_4 + 640.8u_5] - [53.4v_1 + 1003.1v_2] \tag{8}$$

Alirezaee and Afsharian [8] ranked all DMUs with two stages approach. Firstly, they ranked all DMUs with different DEA efficiency scores in accordance to the score, and secondly, they ranked the DMUs with the same DEA efficiency scores in accordance to the proposed Balance Index.

5. Result and Discussion

The CCR model and Balance Index in Alirezaee and Afsharian's method were applied to rank 63 selected stocks in construction and material companies listed in Bursa Malaysia. Table 5 shows the efficiency scores for each of companies (DMU) in 2015. According to Table 5, we found that some companies with the same efficiency score. For example, five efficient companies with the efficiency score of 1 and some inefficient companies such as DMU_{23} (HOCK SENG LEE BERHAD) and DMU_{33} (MALAYSIAN RESOURCES) with 0.59 efficiency score, while DMU_{20} (HEVEABOARD BHD) and DMU_8 (BREM HOLDING BERHAD) is 0.52 efficiency score. DMU_{63} (WOODLANDOR HOLDINGS), DMU_{51} (SEACERA TILES BHD), DMU_{24} (HUME INDUSTRIES BHD) and DMU_{22} (HOCK HENG STONE) were others companies which inefficient and have same efficiency score. Therefore, we cannot rank these DMUs using only the efficiency score. Hence, we calculate Balance Index for comparing the competitive profits and reset the ranking process. More negative values in Balance Index indicates that the stock shows good performance and high rank. The quantities of Balance Index for each DMU are shown in Table 6.

.

IOP Conf. Series: Journal of Physics: Conf. Series **892** (2017) 012010 doi:10.1088/1742-6596/892/1/012010

Table 5. Efficiency score for 2015

DMU	Efficiency score	DMU	Efficiency score	DMU	Efficiency score	DMU	Efficiency score
7	1.00	40	0.41	12	0.23	60	0.09
47	1.00	30	0.41	36	0.23	16	0.09
42	1.00	13	0.40	45	0.22	46	0.08
35	1.00	18	0.40	10	0.21	26	0.08
21	1.00	27	0.37	52	0.19	55	0.07
56	0.90	44	0.35	9	0.18	48	0.07
54	0.72	53	0.34	32	0.18	49	0.06
43	0.68	19	0.34	3	0.17	4	0.06
31	0.67	15	0.32	14	0.16	1	0.05
11	0.65	57	0.32	50	0.16	63	0.03
23	0.59	39	0.31	2	0.15	51	0.03
33	0.59	62	0.29	59	0.14	24	0.03
20	0.52	41	0.29	25	0.14	22	0.03
8	0.52	28	0.27	58	0.14	6	0.02
5	0.50	61	0.24	34	0.13	17	0.00
29	0.48	38	0.23	37	0.11		

According to Table 6, they obtained different negative quantities in Balance Index such as DMU_7 (BOILERMECH HOLD) , DMU_{47} (PINTARAS JAYA BERHAD), DMU_{42} (MULTI-USAGE HOLDINGS), DMU_{35} (MENANG CORPORATION) and DMU_{21}(HO HUP CONSTRUCTION) whose the efficiency score are 1 which is efficient and the quantities Balance Index are -8398.9, -7770.8, -307.5, -211.6 and -146.9 respectively. Based on Table 7, we can see DMU_7 (BOILERMECH HOLD) obtained the top rank and DMU_{21} (HO HUP CONSTRUCTION) was at the bottom rank among efficient companies.

While DMU_{59} (WTK HOLDINGS BHD) , DMU_{25} (IJM CORPORATION BHD) and DMU_{58} (UNITED ULI CORP) are inefficient companies which they got the same efficiency score 0.14 and the quantities Balance Index are -54.4, -41.0 and -34.8 respectively. According to quantity of Balance Index, DMU_{59} (WTK HOLDINGS BHD) get more negative quantities, therefore DMU_{59} (WTK HOLDINGS BHD) obtained better rank as compare to DMU_{25} (IJM CORPORATION BHD) and DMU_{58} (UNITED ULI CORP).

IOP Conf. Series: Journal of Physics: Conf. Series **892** (2017) 012010 doi:10.1088/1742-6596/892/1/012010

Table 6. Balance Index and Efficiency score for 2015

DMU	Eff. score	Balance Index	DMU	Eff. score	Balance Index	DMU	Eff. score	Balance Index	DMU	Eff. score	Balance Index
7	1.00	-8398.9	40	0.41	-60.3	12	0.23	-80.3	60	0.09	-53.9
47	1.00	-7770.8	30	0.41	-83.5	36	0.23	-78.7	16	0.09	-25.3
42	1.00	-307.5	13	0.40	-92.9	45	0.22	-136.2	46	0.08	-46.4
35	1.00	-211.6	18	0.40	-134.0	10	0.21	-127.1	26	0.08	-37.9
21	1.00	-146.9	27	0.37	-97.2	52	0.19	-84.3	55	0.07	-45.2
56	0.90	-165.6	44	0.35	-7485	9	0.18	-37.3	48	0.07	-70.1
54	0.72	-46.0	53	0.34	-250.1	32	0.18	-28.9	49	0.06	-115.2
43	0.68	-228.1	19	0.34	-55.7	3	0.17	-104.8	4	0.06	-63.7
31	0.67	-158.9	15	0.32	-114.2	14	0.16	-66.7	1	0.05	-40.5
11	0.65	-120.3	57	0.32	-66.2	50	0.16	-122.1	63	0.03	-70.8
23	0.59	-15062	39	0.31	-93.8	2	0.15	-118.1	51	0.03	-38.8
33	0.59	-151.6	62	0.29	-94.3	59	0.14	-54.4	24	0.03	-7.2
20	0.52	-129.8	41	0.29	-73.2	25	0.14	-41.0	22	0.03	-37.2
8	0.52	-115.9	28	0.27	-189.3	58	0.14	-34.8	6	0.02	-58.9
5	0.50	-297.9	61	0.24	-63.1	34	0.13	-96.7	17	0.00	-30.5
29	0.48	-94.6	38	0.23	-126.7	37	0.11	-169.2			

Table 7 shows complete ranking of 63 companies. DMU_7 (BOILERMECH HOLD) obtained the highest rank and DMU17 (FAJARBARU BUILD) was the least rank among all DMUs. According to Table 7, we can see ranking of the same efficiency score of DMUs, for example five efficient DMUs which were ranked using their Balance Index or many inefficient DMUs which their efficiency score unit were also the same could be ranked using their Balance Index.

Table 7. Complete ranking for 2015

DMU	Rank	DMU	Rank	DMU	Rank	DMU	Rank
7	1	40	17	12	33	60	49
47	2	30	18	36	34	16	50
42	3	13	19	45	35	46	51
35	4	18	20	10	36	26	52
21	5	27	21	52	37	55	53
56	6	44	22	9	38	48	54
54	7	53	23	32	39	49	55
43	8	19	24	3	40	4	56
31	9	15	25	14	41	1	57
11	10	57	26	50	42	63	58
23	11	39	27	2	43	51	59
33	12	62	28	59	44	24	60
20	13	41	29	25	45	22	61
8	14	28	30	58	46	6	62
5	15	61	31	34	47	17	63
29	16	38	32	37	48		

IOP Conf. Series: Journal of Physics: Conf. Series **892** (2017) 012010 doi:10.1088/1742-6596/892/1/012010

6. Conclusion

This paper also illustrated method of selection and clarify inputs and outputs based on a Fuzzy Delphi and Expert Information. It also shows the efficiency measurement using shadow prices and ranking of stock's companies by using efficiency score. Moreover, this study also demonstrated an empirical problem in complete ranking that were due to some decision making units with the same efficiency score. Decision on the selection the inputs and outputs can affect the ranking of performance evaluation of companies. Consequently, we can rank all of companies to which is efficient or inefficient DMUs easily and completely. In conclusion, the ranking by this method was employed easily and with logical computation.

Appendix

Company's listed in Bursa Malaysia (Construction and Material Industry)

DMU	Companies	DMUS	Companies
1	AHMAD ZAKI RES	34	MELATI EHSAN HOLD
2	AJIYA BERHAD	35	MENANG CORPORATION
3	ASTINO BERHAD	36	MERCURY INDUSTRIES
4	AYS VENTURES	37	MERGE ENERGY BERHAD
5	BINA DARULAMAN BHD	38	MIECO CHIPBOARD
6	BINA PURI HOLDINGS	39	MITRAJAYA HOLDINGS
7	BOILERMECH HOLD	40	ML GLOBAL BHD
8	BREM HOLDING BERHAD	41	MUHIBBAH ENGINEERING
9	CAHYA MATA SARAWAK	42	MULTI-USAGE HOLDINGS
10	CREST BUILDER HLDGS	43	NAIM HOLDINGS BERHAD
11	DKLS INDUSTRIES	44	OKA CORPORATION BHD
12	DOMINANT ENTERPRISE	45	PANSAR BHD
13	ECONPILE	46	PESONA METRO
14	EKOVEST BERHAD	47	PINTARAS JAYA BERHAD
15	ENGTEX GROUP BHD	48	PLB ENGINEERING
16	EVERGREEN FIBREBOARD	49	PRINSIPTEK CORP BHD
17	FAJARBARU BUILD	50	RESINTECH BHD
18	GADANG HOLDINGS	51	SEACERA TILES BHD
19	GAMUDA BERHAD	52	SUPERLON HOLDINGS
20	HEVEABOARD BHD	53	SYCAL VENTURES
21	HO HUP CONSTRUCTION	54	TASEK CORPORATION
22	HOCK HENG STONE	55	TRC SYNERGY BHD
23	HOCK SENG LEE BERHAD	56	TRIPLC BHD
24	HUME INDUSTRIES BHD	57	UEM EDGENTA BHD
25	IJM CORPORATION BHD	58	UNITED ULI CORP
26	JAKS RESRCS BHD	59	W T K HOLDINGS BHD
27	KERJAYA PROSPEK	60	WCE HOLDINGS BHD
28	KIA LIM BERHAD	61	WCT HOLDINGS BHD
29	KIM HIN INDUSTRY BHD	62	WHITE HORSE BERHAD
30	KIMLUN CORPORATION	63	WOODLANDOR HOLDINGS
31	KPS CONSORTIUM BHD		
32	LAFARGE MALAYSIA BHD		
33	MALAYSIAN RESOURCES		

References

[1] A. Charnes, W. W. Cooper, and E. Rhodes Measuring the efficiency of decision making units 1978 *Eur. J. Oper. Res.* **2** 6 429–444
[2] G. R. Jahanshahloo and P. Firoozi Shahmirzadi New methods for ranking decision making units based on the dispersion of weights and Norm 1 in Data Envelopment Analysis 2013 *Comput. Ind. Eng.* **65** 2 187–193
[3] F. Yang, S. Ang, Q. Xia, and C. Yang Ranking DMUs by using interval DEA cross efficiency matrix with acceptability analysis 2012 *Eur. J. Oper. Res.* **223** 2 483–488
[4] P. Andersen and N. C. Petersen A Procedure for Ranking Efficient Units in Data Envelopment Analysis 1993 *J. Manag. Sci.* **39** 10 1261–64
[5] M. Khodabakhshi and K. Aryavash Ranking all units in data envelopment analysis 2012 *Appl. Math. Lett.* **25** 12 2066–70
[6] F. Hosseinzadeh Lotfi, A. A. Noora, G. R. Jahanshahloo, and M. Reshadi One DEA ranking method based on applying aggregate units 2011 *Expert Syst. Appl.* **38** 10 13468–71
[7] F. Hosseinzadeh Lotfi, R. Fallahnejad, and N. Navidi Ranking efficient units in DEA by using TOPSIS method 2011 *Appl. Math. Sci.* **5** 17–20 805–815
[8] M. Afsharian and M.R. Alirezaee A complete ranking of DMUs using ristriction in DEA models 2007 *Appl. Math. Comput.* **189** 1550–59
[9] D. Guo and J. Wu A complete ranking of DMUs with undesirable outputs using restrictions in DEA models 2013 *Math. Comput. Model.* **58** 5 1102–09
[10] J. Wu, F. Yang, and L. Liang A modified complete ranking of DMUs using restrictions in DEA models 2010 *Appl. Math. Comput.* **217** 2 745–751
[11] R. Arsad, M. N. Abdullah, and S. Alias A ranking efficiency unit by restrictions using DEA models 2014 *AIP Conf. Proc.* vol 1635 pp 266–273
[12] M. Mokhtar, A. Shuib, and D. Mohamad Identifying the critical financial ratios for stocks evaluation: A fuzzy delphi approach 2014 *AIP Conf. Proc.* vol 348 pp 348–354
[13] B. K. Sahoo and E. Meera A Comparative application of alternative DEA models in selecting efficient large cap market securities in India 2008 *Int. J. Manag. Perspect.* **1** 2 62–75
[14] A. Mansouri, N. Ebrahimi, and M. Ramazani Ranking of companies based on TOPSIS-DEA approach methods (case study of cement industry in Tehran stock exchange) 2014 *Espacios* **35** 7 10
[15] J. Powers and P. McMullen Using Data Envelopment Analysis to select efficient large market cap securities 2000 *J. Bus. Manag.* **7** 2 31–42
[16] A. Ana, L. Miranda, M. Vinicius, A. De Lima, A. Dutra, and E. A. Lanzer DEA Investment strategy in the Brazilian stock market 2007 *EnANPAD* **13** 1–10
[17] O. P. Ling and A. A. Kamil Data Envelopment Analysis for stocks selection on Bursa Malaysia 2010 *Arch. Appl. Sci. Res.* **2** 5 11–35
[18] N. C. P. Edirisinghe and X. Zhang Input / output selection in DEA under expert information with application to financial markets 2010 *Eur. J. Oper. Res.* **207** 3 1669–78

Towards an Enhancement of Organizational Information Security through Threat Factor Profiling (TFP) Model

Fatimah Sidi[1], Maslina Daud[2], Sabariah Ahmad[2], Naqliyah Zainuddin[2], Syafiqa Anneisa Abdullah[2], Marzanah A.Jabar[1], Lilly Suriani Affendey[1], Iskandar Ishak[1], Nurfadhlina Mohd Sharef[1], Maslina Zolkepli[1], Fatin Nur Majdina Nordin[1], Hashimah Amat Sejani[3], Saiful Ramadzan Hairani[3]

[1]Faculty of Computer Science and Information Technology, University Putra Malaysia, 43400 UPM Serdang, Selangor Darul Ehsan, Malaysia.

[2]CyberSecurity Malaysia, Level 7, SAPURA@MINES, No. 7, Jalan Tasik, The Mines Resort City, 43300, Seri Kembangan, Selangor, Malaysia.

[3]InfoComm Development Centre (iDEC), University Putra Malaysia, 43400 UPM Serdang, Selangor Darul Ehsan, Malaysia.

fatimah@upm.edu.my

Abstract. Information security has been identified by organizations as part of internal operations that need to be well implemented and protected. This is because each day the organizations face a high probability of increase of threats to their networks and services that will lead to information security issues. Thus, effective information security management is required in order to protect their information assets. Threat profiling is a method that can be used by an organization to address the security challenges. Threat profiling allows analysts to understand and organize intelligent information related to threat groups. This paper presents a comparative analysis that was conducted to study the existing threat profiling models. It was found that existing threat models were constructed based on specific objectives, thus each model is limited to only certain components or factors such as assets, threat sources, countermeasures, threat agents, threat outcomes and threat actors. It is suggested that threat profiling can be improved by the combination of components found in each existing threat profiling model/framework. The proposed model can be used by an organization in executing a proactive approach to incident management.

1. Introduction

Information security activity is defined as "to apply any technical methods and managerial processes on the information resources (hardware, software and data) in order to keep organizational assets and personal privacy protected" [1]. Meanwhile, security risk is defined as the harm to a process or the related information resulting from some purposely or accidental event that negatively impacts the process or the related information [2, 3].

Recently, organizations have identified information security as an area of their operations that requires to be protected and to be integrated into their internal control system. This is because information, or also known as information asset, is a valuable asset that should be appropriately protected by an organization. Information assets are is differently defined by organizations as being information or data that is of value to the organization, such as patient records, intellectual property, or customer information [1, 4].

In addition, the high reliance of an organization to devices or computers and the engagement with Internet services forces organizations to seriously consider the importance of security issues [5]. Statistics reported by the Malaysian Computer Emergency Response Team (MyCERT) show that incidents of fraud and intrusion are among the top three ranked incidents in Malaysia since 2014 [6]. Thus, effective information security management is required, as information security is facing unprecedented challenges due to the serious threat from unauthorized users on the Internet [2].

Normally, threats are not to be viewed as single events. But threats are often related to a number of other threat scenarios that form a campaign of threats and attacks against organizations [7]. The threat could be from automated input, environmental input or situational information, which lead to a bigger risk if it is not managed appropriately. Thus, threat factor profiling is a method used to help an organization to address the security challenges [4].

Threat factor profiling will provide organizations with proactive approach for incident management by planning, training and developing risk-mitigating strategy. This is because threat factor profiling is an analytics technique that could assist network security analysts to understand and organize intelligence information related to threat groups. Based on research and study conducted in this area, organizations in Malaysia are lacking on threat profiling and metric in managing information security risks and current threats.

Hence, this paper aims to propose a Threat Factor Profiling Model for Information Security. This model was adapted from a comparative analysis on existing information security threat approaches, and based on combination of component from reviewed models [6]. In managing information security, it is vital to identify the factors that may lead to a failure in the confidentiality, integrity or availability (CIA) of an information system. Therefore, a threat profile of an organization should include all threat information such as critical assets, threat actors and threat scenarios. We can also refer to the elements of risk management such as assets, threats, vulnerabilities and controls, which are jointly evaluated with the variables of probability and impact [6]. In this paper, we describe the main components or factor of the proposed Threat Factor Profiling Model.

The rest of the paper is organized as follows: section 2 provides a review of our previous work on comparative analysis of existing model/framework, section 3 provides justification of the proposed threat model, section 4 describes the characteristic, threat modeler and components of threat profiles, and section 5 offers discussion and conclusion for the paper.

2. Previous Work

In developing a threat factor profiling model, an organization must determine the information assets that must be protected and the degree of protection to be provided for them [8]. This is because a threat will exploit the vulnerability of an asset or group of assets and thereby cause harm to the organization. Each day, organizations face an increase of threats to their networks and the services that will lead to information security issues [5].

The goals and outcome of a threat factor profiling model need to be quantified in order to understand the potential attack vectors and encounter them. Therefore, the steps in developing the Threat Factor Profiling Model includes literature review and comparative analysis of existing models, followed by a model proposal. This paper is an extended work from our previous work, which is a comparative analysis of existing model/framework [6].

Threat modelling is an example of approach, method, framework and technique used in analyzing security of an application [9]. In addition, numerous standards are available as reference in risk management. The standards specify and point out the stages of identification in risk management process [10]. Many organizations using threat modelling technique and risk management standard to secure their cyber network to prevent financial loss, as well as to handle threats in their organizations [6].

IOP Conf. Series: Journal of Physics: Conf. Series **892** (2017) 012011 doi:10.1088/1742-6596/892/1/012011

Based on our previous comparative analysis, we found that each model namely STRIDE [11], DREAD [12], OCTAVE [13], CVSS [14], VERIS [15], AS/NZS 3100:2009 [16], ISO/IEC 27005 [17, 18] and NIST SP 800-39 [19] have different approaches, components and objectives. The key components identified from existing models were assets, threat sources, countermeasures, threat agent, threat outcome and threat actors [6]. However, it is suggested that a better threat profiling model to be introduced. The threat factor profiling model could identify and mitigate risks by combining the components found in each existing threat model.

Our previous comparative analysis discovered that STRIDE and CVSS were capturing a major share of attention in terms of threat identification and management for organizations [8]. STRIDE is used to develop an understanding of risks to a system and how to mitigate them. CVSS is a comprehensive scoring and assessment risk model that provides standardized vulnerability scores.

STRIDE is more advanced than other models in terms of processes because it includes the process of measuring threats, which is not available in other models [20]. STRIDE can identify threat sources and the identified threat can be categorized based on the criteria [21]. The key components for STRIDE model are asset, threat source and countermeasure [6].

CVSS is an open framework for communicating the characteristics and severity of software vulnerabilities. CVSS uses the numerical approach to identify threat by calculating the impact and requires a spreadsheet to calculate the risk component [22]. Therefore, the key components for this framework are asset and threat source [6].

AS/NZ model is another formal standard used in documenting and managing risk, not specifically for cyber threats. AS/NZ model provides a common approach in support of standards dealing with specific risks and/or sectors [24]. The key component for this model is asset [6].

Open Web Application Security Project (OWASP) model is a standard for secure web development and testing as it provides a guide to define security requirements to build secure web applications [23]. It is used as a guidance for web application security project development.

Another important element in developing a threat profiling model is threat metric as it can be calculated, and the severity of the threat level to reduce the threats can be acquired. Metric is defined as a measurement, taken over a period of time that communicates vital information about a process or activity. Choosing a proper metrics can provide valuable insight and control into our systems to ensure they are secure in a very cost effective manner [25].

However the threat metric element has not been emphasized in any existing models. In fact, the derived threat metric would enable suitable actions to be taken based on the level of severity and vulnerability to the organization.

It is also discovered that the generic threat profile lacks the component to generate important information for strategic decision making [6, 7]. Therefore the proposed model will include the strategic decision component by incorporating the threat metrics. The vulnerabilities and risk components are unified in the threat scoring of the proposed model.

Thus, the adopted components from the existing models are (i) threat sources, (ii) threat motive, (iii) threat outcomes, (iv) threat agents and (v) threat. The details will be elaborated in the next sections.

3. Justification

Based on the above literature review, existing mechanism used in handling threat within an organization are threat modelling techniques, framework and methods, risk management standards, and many other threat approaches. These mechanisms have the same purpose, which are to combat multiple security issues and protecting the organization's information assets. However, each mechanisms, models, framework and standards are restricted to some parts of the security component, and is not applicable to all information security risk management activities. Besides, none of the models are adaptable to highly dynamic environments.

Due to the current challenges and rapid technological development, organizations face more sophisticated and prevalent advanced persistent threat (APT) scenarios that could not be handled by existing security risk management programs. The programs are not able to evaluate the security risks completely and the incident management programs are not able to defend against the threat [7]. Therefore, the Threat Factor Profiling for Information Security is proposed as it integrates all the key components of security threat profiling towards producing a comprehensive and dynamic threat model. Here we highlight some of the advantages of the proposed model compared to the existing model/framework;

1. Determination of threat sources.
2. Categorization and differentiation of threats.
3. Determination of severity level.
4. Identification of related threat scenarios that constitute the full threat scenario campaign.

4. Proposed Threat Factor Profiling Model

4.1. Characteristic of the proposed threat model

The four main characteristics identified for the proposed model were adopted from the existing models. First, the proposed threat factor profiling model is based on security system standards based on CIA [6]. Second, the threat categorization in this proposed model is based on the STRIDE model. Third, the risk assessment approach in the proposed threat model is based on AS/NZ model. Fourth, the impact and threat scoring that is implemented in the proposed model is based on the CVSS model.

4.2. Threat profiling model

Figure 1 describes the proposed Threat Factor Profiling Model for Information Security. The main components in this threat profile are input sources, threat agents that include threat motives, threat actors, and threat outcomes, followed by threat metric, as well as scoring engine and dashboard visualization.

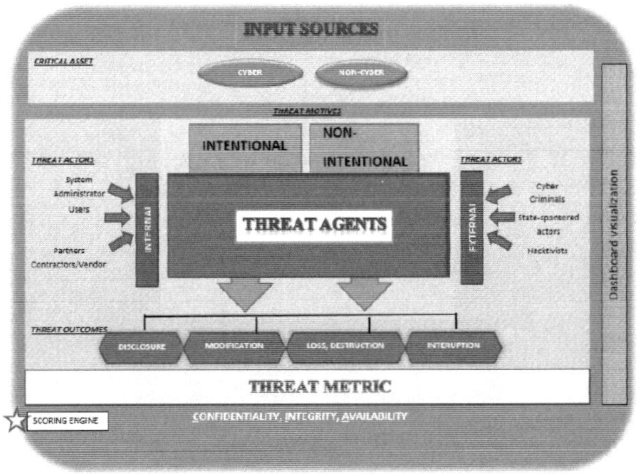

Figure 1. Proposed Threat Factor Profiling Model for Information Security

Figure 2 depicts the flow for generating scenario of a threat profile. The flows are divided into three main parts, namely, (i) Threat Sources, (ii) Threat Profile that include Scoring and Threat Metric, and (iii) Threat Profile Report.

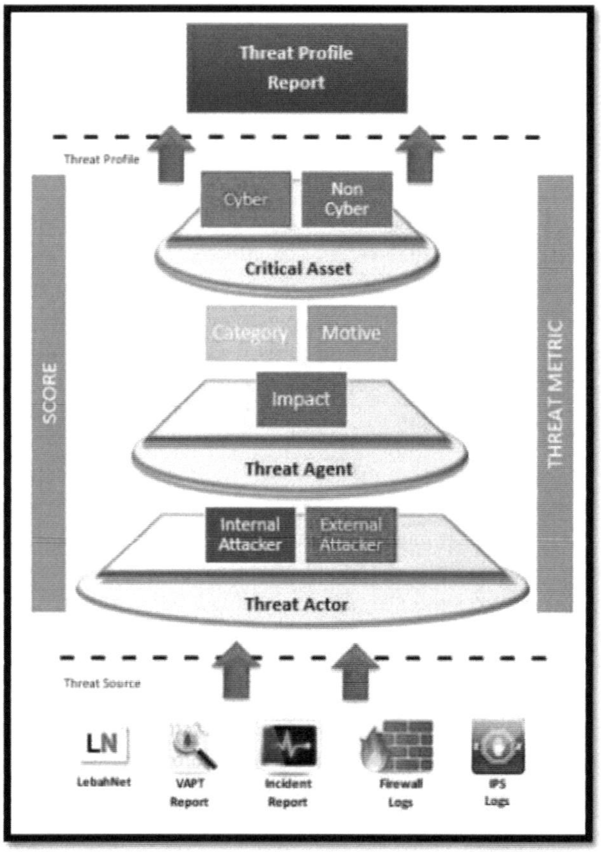

Figure 2. Flow for generating scenario of a threat profile

(i) Threat Sources
Threat is defined as an event that could have some type of negative impact on the organization. Possible threat sources in this project come from internet and intranet such as sensor, Vulnerability Assessment and Penetration Test (VAPT) report, incident report, firewall logs and IPS logs. Threat from various sources shall be analyzed from the network sensor.

(ii) Threat Profile
A threat profile includes information about threat actor, threat agent, critical asset, and threat scenarios. A threat scenario is an illustration in which one or more threat actors can mount one or more threat actions in an attempt to compromise an identified critical asset by exploiting both vulnerabilities and inadequate safeguards [24]. A threat scenario campaign is a series of related threat scenarios that are used together as part of an advanced persistent threat (APT) for a common objective.

a) Scoring
The assessment of threat impact shall require a scoring mechanism to indicate the severity level. CVSS will be adapted to compute the score.

b) Threat Metric
Threat metrics shall be produced based on attributes from threat scenarios, threats, critical assets and others. These threat metrics will be used for the generation of the threat matrix. Threat matrix is used to characterized and differentiate threats. Together with threat scoring, the threat matrix could generate strategic information.

IOP Conf. Series: Journal of Physics: Conf. Series **892** (2017) 012011 doi:10.1088/1742-6596/892/1/012011

(iii) Threat Profile Report
Threat profile report is the final step in this proposed model. Threat profile is produced based on the collection of scenarios that shall form the threat matrix. The threat ranking is produced using threat scoring together with the threat matrix and expert judgement/rules from the security analyst. It is used to determine the severity level. The threat profile shows the pattern of threat scenarios that threaten the critical assets, the resulting impact, and countermeasure for the organization. Together, they provide a picture of the security risk that the organization is facing.

4.3. Details of components in Threat Profile
The main components identified for Threat Profile were; (i) Critical Asset, (ii) Threat Agent, and (iii) Threat Actor. All components were found to be important in identifying threat and its vulnerability.

(i) Critical Asset
Threat victim asset is something of value to the organization (information in electronic or physical form, information systems, and a group of people with unique expertise). The examples are information technology assets that are commonly compromised and used during attacks include, but are not limited to, servers, network components, user devices, storage media, people, network and system design specifications, and Virtual Private Network (VPN) configurations. Critical information assets, which are usually the final objective of the threat actor, include intellectual property, product development information, manufacturing processes, business plans, policies, emails, organization charts, and user credentials. It also can be divided into cyber and non-cyber categories. Organizations need to evaluate the impact of threats to critical assets (based on NICCS glossary [26]).

(ii) Threat Actor
Threat actor refers to whom or what may violate the security requirements (confidentiality, integrity, availability- CIA) of an asset. Actors can be from inside or outside the organization. There are two type of threat actors, which refer to inside actor and outside actor. Internal attacker refers to system administrator, end users, executives and partners. External attacker refers to cybercriminal and hacker (based on NICCS glossary [26]).

(iii) Threat Agent
Threat agents there will be divided into category, motive, impact and score. The term threat agent is used to indicate an individual or group that can manifest a threat. It is fundamental to identify who would want to exploit the assets of a company, and how they might use them against the company. Threat agents can be classified into 3 as follows:

a) Category (based on British Standard Institute -BSI Standards) including Force Majure, Deliberate Act, Human Failure, and Technical Failure,
b) Motive (based on VERIS Standards [13]) that can be deliberate (intentional) or accidental (non-intentional), and
c) Impact (Outcome) is the immediate result of violating the security requirements of an asset. The resulting outcome or effect of threat scenarios typically falls into these categories (based on OCTAVE [11]).

5. Discussion and Future Works
The paper presents a proposed Threat Factor Profiling Model for Information Security that is adapted from existing models/frameworks and has been refined to allow decision making capability. The threat factor profiling model is an integrated model as it combines the key components or factors found in

existing models. As mentioned by Irwin [7], an organization's threat profile should include all threat information and present a clear and detailed illustration of how each of the components are used together. Thus, the threat factor profiling model shall be able to provide a clear illustration of threat and able to help an organization in executing a proactive approach to incident management.

6. References

[1] Hong KS, Chi YP, R.Chao L. and Tang JH 2003 An integrated system theory of information security management *Information Management And Computer Security* **11(5)** pp 243-248
[2] Schultz EE, Proctor RW and Lien MC 2001 Usability and security: an appraisal of usability issues in information security methods *Computer & Security* **20(18)** pp. 620-634
[3] Elky S 2006 An Introduction to information system risk management *SANS Institute* Available at https://www.sans.org/reading-room/whitepapers/auditing/introduction-information-system-risk-management-1204 [Accessed: 2-Feb-2017]
[4] Stevens JF 2005 Information Asset Profiling *Technical Note* (Carnegie Mellon University June 2005) Available at https://resources.sei.cmu.edu/asset_files/TechnicalNote/2005_004_001_14507.pdf [Accessed: 2-Feb-2017]
[5] Zulkernine SI, Mohammad and Ahamed 2006 Software security engineering: toward unifying sofware engineering and security engineering *Enterp. Inf. Syst. Assur. Syst. Secur. Manag. Tech. Issues* M.Warkentin & R.Vaughn, pp.215-232
[6] Fatimah Sidi, A.Jabar Marzanah, Lilly Suriani Affendey, Iskandar Ishak, Nurfadhlina Mohd Sharef, Maslina Zolkepli, Tan Ming Ming, Muhammad Faidhi Abd Mokhti, Maslina Daud, Naqliyah Zainuddin and Rafidah Abdul Hamid 2017 A Comparative Analysis Study on Information Security Threat Models: A Propose for Threat Factor Profiling. *Journal of Engineering and Applied Sciences*, **12** 548-554. DOI: 10.3923/jeasci.2017.548.554
[7] Irwin S, 2014 Creating a threat profile for your organization *The SANS Institute* pp. 1-31. Available at https://www.sans.org/reading-room/whitepapers/threats/creating-threat-profile-organization-35492 [Accessed: 2-Feb-2017]
[8] Dutta A. and McCrohan K 2002 Management role's in information security in a cyber economy *California Management Review* **45(1)** pp. 67-87 DOI: 10.2307/41166154
[9] Kadhirvelan SP and Soderberg-Rivkin A 2014 Threat modelling and risk assessment within vehicular systems (Chalmers University of Technology August 2014). Available at http://publications.lib.chalmers.se/records/fulltext/202917/202917.pdf [Accessed: 2-Feb-2017]
[10] Zhang XZX, Wuwong N, Li HLH and Zhang XZX 2010 Information security risk management framework for the cloud computing environments *Computer and Information Technology (CIT)* (2010 IEEE 10th International Conference) pp. 1328-1334 DOI: 10.1109/CIT.2010.501
[11] Scandariato R, Wuyts K and Joosen W 2015 A desriptive study of Microsoft's threat modelling technique *Requir. Eng.* **20(2)** pp. 163-180.
[12] Rao KRM and Pant D 2010 A threat risk modelling framework for Geospatial Weather Information System (GWIS): A DREAD based study *IJACSA Int. J.Adv.Comput. Sci.Appl.* **1(3)** pp. 20-28
[13] Feng N, Wang HJ and Li M 2014 A security risk analysis model for information system: Causal relationships of risk factors and vulnerability propagation analysis *Inf. Sci. (Ny)* **256** pp. 57-73
[14] Ibidapo AO, Zavarsky P, Lindskog D, and Ruhl R. 2011 An analysis of CVSS v2 environmental scoring *2011 IEEE International Conf. Privacy, Secur. Risk Trust IEEE Int. Conf. Soc. Comput. PASSAT/SocialCom 2011 – Proc* pp.1125-1130
[15] Obrst L, Chase P and Markeloff R 2012 Developing an ontology of the cybersecurity domain *STIDS* pp. 49-56.

[16] Hussain S, Kamal A, Ahmad S, Rasool G, and Iqbal S 2014 Threat modelling methodologiea: A survey *Sci. Int. (Lahore)* **26(4)** pp. 1607-1609
[17] Lalanne V, Munier M and Gabillion A 2013 Information security risk management in a world of services *2013 Int. Conf. Soc. Comput.* pp. 586-593
[18] Leitner A. and Schaumuller-Bichl I 2009 Arima-a new approach to implement ISO/IEC 27005 *2009 2nd Int. Symp. Logist. Ind. Informatics, LINDI* pp. 1-6.
[19] J.T.F.T. Initiative 2011 Managing information security risk *Nist. Spec. Publ. March* pp. 320
[20] Sultan R. and Abbas SQ. 2015 Web Services Threats, Vulnerabilities and Countermeasures *Int. J. Adv. Res. Comput. Sci. Manag. Stud.* **3(3)** pp. 243-252
[21] Xin T. and Xiaofang B 2014 Online banking security analysis based on STRIDE threat model *Int. Journal of Security and its Apllications* **8(2)** pp. 271-282
[22] Gallon L 2011 Vulnerability discrimination using CVSS framework *2011 4th IFIP Int. Conf. New Technol. Mobil. Secur. NTMS 2011 – Proc.*, pp.1-6
[23] Vibhandik R. and Bose A.K 2015 Vulnerability assessment of web applications – A testing approach *IEEE* pp. 16-21 ISBN 978-1-4799-8451-1/15
[24] Sheddan P, Ruighaver T. and Atif A. 2010 Risk management standards – the perception of ease of use J.Int. Sysy. Secur. **6(3)** pp. 1-13
[25] Cambra R. 2004 Metrics for operational security control GIAC Security Essentials Certification (GSEC) – Practical Assignment, pp. 1-15.
[26] Glossary of NICCS. Availabe at https://niccs.us-cert.gov/glossary#C. [Accessed: 2-Feb-2017].

Acknowledgement
This research is supported by the Ministry of Science, Technology and Innovation (MOSTI) under a special grant scheme programme **The National Policy on Science, Technology & Innovation (**DSTIN) **Flagship Programme.** This research is a collaboration work between CyberSecurity Malaysia and University Putra Malaysia (UPM) to jointly develop the National Integrated Information Security Threat Profiling Model (NIISTFP). Any opinions, finding, and conclusions or recommendations expressed in this material are those of the authors and do not necessarily reflect the view of MOSTI.

Appropriation of sociaal media for fostering effective tacit knowledge sharing: developing conceptual model

A Amidi, M Jabar, Y Y Jusoh and R Abdullah

[1]Department of software engineering and information system , Faculty of Computer Science and Information Technology, University Putra Malaysia, 43400 UPM Serdang, Malaysia

marzanah@upm.edu.my

Abstract. With the rising popularity of social media in the past few years, several researches ratiocinate that this type of interactive and collaborative technology could be a beneficial tool for the sharing of tacit knowledge. Nevertheless, very few literatures have tackled the subject of how social media could facilitate tacit knowledge sharing among medical practitioners, and what are its contributions in the area. Thus, the factors that drive individuals to share tacit knowledge need to be investigated further and included in literature. Through a systematic literature review, this study proposes seven enabling conditions which could potentially facilitate the sharing of tacit knowledge. TAM was applied as a novelty in this study in investigating the factors influencing knowledge sharing via social media, whilst taking into account the mediation effects of Attitude in social media usage. This study uncovered an important correlation between virtual settings and the conversion of tacit knowledge, which affects organizational members who are not co-located physically but have a crucial need for sharing information.

1. Introduction

Knowledge management (KM) has long been the main focus on social media for the past decade. Computer-mediated communications have come to depend on Online Social Networks (OSNs) as a crucial social platform KM initiatives involving information technology have been implemented by various organizations who have also invested heavily in the deployment of their KM systems. Such measures enable their employees to share explicit knowledge among themselves through corporate knowledge repository. Social media contributes new ways of managing knowledge at personal and organizational levels via social-collaborations and networking opportunities. Nonaka [1] explain two types of knowledge tacit and explicit. Tacit knowledge involves the internalization of individual processes such as experience, reflection, internalization or individual talents, which cannot be managed and learned in the same way as explicit knowledge. Tacit knowledge is stored exclusively within the human self, whilst explicit knowledge can be technologically or mechanically deposited in the form of information systems or handbooks for example. Since tacit knowledge is fundamentally technical or cognitive consisting of mental models, values, beliefs, perceptions, insights and assumptions [2], it is thus complicated to be communicated or documented in a form that could be shared with others. On the other hand, based on most knowledge management (KM) frameworks, explicit knowledge can be managed through the processes of capturing, storing, retrieving, transferring and applying. This indicates that tacit knowledge is not only an important organizational resource, but also a strategic one as it is the only form of information that is renewable and sustainable which is an important competitive advantage for any organization. Tacit knowledge is estimated to contribute up to 80% of vital knowledge, which could prove to be beneficial in ensuring an organization's sustainability in competitive markets [3].

A review of literature shows a gap in the studying of the challenges involved in the establishment of social media for the effective and efficient sharing of tacit knowledge. Various discussions have been made with regards to the transferability of tacit knowledge and the best methods for transferring it. Despite the introduction of numerous models for tacit knowledge transfer, none have been specifically established. Although tacit knowledge transfer has been modeled fairly well in theory, practical problems still exist due to the strong linkage between this type of knowledge with the owner of the knowledge and his personality. Several researchers view social web tools as enablers of tacit knowledge sharing [4-13]. Their arguments indicated that certain technological factors have an influence on these variables with regards to knowledge sharing via the platform of social media. Although these researchers offer a view into the affordance of specific factors that influence knowledge sharing via social media, they do not particularly aim to comprehend these affordances for tacit knowledge sharing. Hence, the first main issue being tackled by this research concerns the viability of ICT in general for the sharing of tacit knowledge [4, 14, 15]. This study aims to close this gap by studying the motivational factors that facilitate the sharing of tacit knowledge. Hence, the proposed research model draws together technological factors that affect tacit knowledge sharing.

2. Theoretical background

In social media literature, the terms 'online communities', 'social media', 'virtual communities' and 'online social groups' are used interchangeably. Social media platforms are defined as a set of internet based tools that facilitate the creation and exchange of user generated content These platforms consist of blogs, video-sharing sites, micro-blogs, forums, websites and wikis that are developed based on the concepts of Web 2.0 The digitization of healthcare is addressed by Stewart et al [8] who also examined the opportunities for IS research in the context of the healthcare industry. The researchers highlighted non-traditional areas such as the appropriation of social media by physicians, with a provision of opportunities for IS experts in this area through the creation of physician-centered virtual communities The study also indicated that another potential research area for IS experts lies in issues related to the adoption and use of social media [16,17,60]. Another research conducted in Quebec, Canada indicated that female health professionals viewed Web 2.0 as a useful mechanism for transferring knowledge, but they also noted lack of time and technological skills as a significant limitation for them [18]. A social network analysis investigating the knowledge-sharing behavior among practitioners in an online clinical forum discovered that despite their limited number, the practitioners' inter-professional and inter-institutional links are strong [8]. Gormley [19] attempted to manage knowledge resources through the offering of a social media-based knowledge-sharing opportunity in the form of virtual communities of practice (VCoPs) to employees who are involved in strategic areas of the healthcare sector and its delivery.

The healthcare sector is comprised mostly of tacit knowledge [14], which forms 80% of the overall vital knowledge [3] that is necessary for better performance [20]. Although tacit knowledge amongst healthcare professionals is a crucial source of experiential know-how, numerous operational and technical reasons have rendered it to be unharnessed and unapplied in practice [21]. Tacit healthcare knowledge could contribute significantly to the enhancement of healthcare quality and delivery, when applied together with explicit (published) healthcare knowledge. Regular ICT tools such as emails and databases lack the appropriate setting for tacit knowledge transfer. Considering that the sharing of explicit knowledge renders less effort than tacit knowledge sharing [22], there is therefore a requirement for more open and unstructured solutions [23]. With the growing popularity of social media, healthcare educators need to think of different ways to communicate [24] considering that the utilization of social media in the healthcare sector is still new with undiscovered depths and only partially realized advantages [25].

IS research on system appropriation normally tackles the common aspects of ease of use usefulness and perceived enjoyment as employed by theories like TAM in the domain of knowledge sharing of which emphasis is on the sharing of explicit knowledge. Collaborative work can be facilitated by information technology (IT) which can also help enable the process of knowledge sharing [26]. Nevertheless, technologies of such nature have limited abilities in transferring tacit knowledge [27]. As argued by researchers, technical infrastructures rely heavily on the value of the contents that they hold and the relationships that they can build. The motivation to act has been related to two aspects of systems use [28].

Mun, et al. [29] and Kim [30] in their investigation found inter-correlations between technology acceptance and knowledge sharing participation. In short, the knowledge sharing level amongst employees could indicate in their intention in using Social media. This study defines knowledge sharing as the extent of information and knowledge sharing amongst employees in an organization. A subsection
Some text.

3. Conceptual model

The TAM Davis [31] was initially used to understand and predict user adoption of new information technologies. TAM, the users' acceptance of new technology is influenced by three major variables: perceived ease of use and perceived usefulness, perceived enjoyment. Perceived ease of use is the extent to which an individual perceives the usage of a certain system to be effortless physically and mentally. Meanwhile, perceived usefulness is the extent to which an individual perceives the usage of a certain system to be beneficial in enhancing his job performance. Knowledge sharing has been found to be driven by perceived ease of technology [32,33]. TAM has recently been employed to investigate the use of web information systems [34] and mobile healthcare systems [35]. Hsu and Lin [36] found that perceived ease of use significantly influences knowledge sharing in blogs. It was also found to encourage information sharing amongst the members of a certain network [37]. Knowledge sharing is more likely to occur when a social media user experiences an ease of interaction with the website. This indicates that the user is more likely to write a review of a product or service if the user perceives the review submission to be easy. Various studies have noted that perceived ease of use has an influence on knowledge sharing behaviors in virtual settings [38]. Similarly, other studies have also found a strong and consistent correlation between perceived usefulness and the use of a certain technology with sharing medical knowledge [39]. Li and Liu [40] found that perceived usefulness positively affects knowledge sharing in online travel services.

Despite all these findings, there is a lack of investigation on the impact of perceived ease of use and perceived usefulness on tacit knowledge sharing in an online setting. Hence, the following hypotheses are extrapolated:

Hypothesis 1: Perceived usefulness positively associate with tacit knowledge sharing in social media.

Hypothesis 2: Perceived ease of use positively associate with tacit knowledge sharing in social media.

Perceived Enjoyment is an addition to the main constructs in TAM that represents intrinsic motivations. Davis, et al. [41] defined it as the degree to which the ICT usage is perceived to be enjoyable. Perceived enjoyment has been deemed as a strong predictor that justifies a certain behavior such as technology adoption and utilization [41-45]. Social media sites commonly focus on utility, which is derived from the sense of entertainment and enjoyment from using them. Wasko and Faraj [46] found empirical evidence that perceived enjoyment has a positive correlation to knowledge contribution in electronic networks of practice.

Meanwhile, Yu, et al. [47], Papadopoulos, et al. [48] and, Kankanhalli, et al. [49] found a positive correlation between perceived enjoyments of blogging with individual attitude towards knowledge sharing. Hau, Kim, Lee and Kim [22], and Kang and Schuett [50] supported the finding that perceived enjoyment enhances experience or knowledge sharing behavior in social media. The sharing of tacit knowledge normally involves intensive exchange processes such as years of intimate conversations on and off work, sharing of experiences and transfer of expertise [51]. Thus, the sharing of tacit knowledge takes more interactions over a longer period of time than explicit knowledge would. Considering that perceived enjoyment in knowledge sharing could be more crucial in the intimate and intense conversational processes among individuals, the hypothesis below is hence posited:

Hypothesis 3: Perceived enjoyment positively associate with tacit knowledge sharing in social media.

Although the topic of knowledge sharing is deemed important in ethics, vague terminologies still characterize discussions related to it. To fill this gap, a framework was developed for this study to better understand how social media affects the sharing of tacit knowledge and particularly to identify the degree of attitude towards social media usage as the mediator in the relationships between perceived ease of use, perceive usefulness, and perceived enjoyment with tacit knowledge sharing. Hence, one of research objectives is to develop a deep understanding of the formation, mediators, and antecedents of knowledge sharing through the opening of a black box of tacit knowledge sharing, which will eventually facilitate the success of tacit knowledge sharing. Technology acceptance is tested either in the form of behavioral intention or actual use. Attitude as a mediator is discarded due to its lack of validation [52, 53]. Attitude towards Using is taken from the construct by Davis et al.[31] derived from the technology acceptance model (TAM) that consists of Perceived Usefulness (PU) and Perceived Ease of Use (PEOU).

As a whole, empirical support from the TAM research leans more towards the importance of PU over PEOU and deems PU as a full mediator for PEOU and acceptance of technology [53,54]. Lin and Lu [55] in their study indicated the significance of perceived enjoyment and perceived usefulness as mediators. McGowan et al. [39] conducted an empirical study among physicians in the USA and found that the use of social media for information and medical knowledge sharing with other physicians relies heavily on the perceived usefulness of the technology and the general attitude that the physicians have towards the value that this technology has to offer. This study has shown that a positive attitude towards the use of social media is more likely to drive the use of social media and the sharing of medical information with other physicians through its use. In short, users will have a higher inclination to view a posted content when they perceive this information to be useful Perceived Enjoyment is also found to have a significant effect on user attitude with regards to using Facebook [56]. Therefore the hypotheses below are posited:

Hypothesis 4: Attitude toward social media usage positively related to the intention tacit knowledge sharing over social media.
Hypothesis 5: Attitude towards social media usage mediates the correlation between perceived enjoyment and the intention of sharing tacit knowledge.
Hypothesis 6: Attitude towards social media usage mediates the correlation between perceived ease of use and the intention of sharing tacit knowledge.
Hypothesis 7: Attitude towards social media usage mediates the correlation between perceived usefulness and the intention of sharing tacit knowledge.

As noted earlier on, TAM has gained significant focus in the field of information system (IS) over the last decade [36]. TAM proposes perceived usefulness, perceived ease-of-use and perceived enjoyment as the three primary factors of IS acceptance in its attempt to predict and explain system use. Attitude towards Using is derived from the construct by Davis et al. [31] from the technology acceptance model (TAM) which consists of Perceived Usefulness and Perceived Ease of Use.

Perceived ease of use is the degree to which a user perceives the usage of the technology to be effortless. In this study, perceived ease of use refers to the extent to which an individual perceives the usage of social media to post or retrieve medical information to be effortless. Meanwhile, perceived usefulness is the degree to which an individual perceives that the use of the technology will improve his job performance [31]. This in turn will increase the individual's willingness to use it further in the future [57]. This model posits that perceived usefulness and perceived ease-of-use affect attitude towards usage. When a user perceives that the available tools could actually facilitate in a certain task's completion, the user will acknowledge the significance of the tool and thus make full use of it. Other studies suggest that perceived enjoyment the extent to which the user perceives the technology to be enjoyable, could also influence user attitude towards using the technology. Venkatesh et al. [58] found that there is a positive link between perceived enjoyment, perceived ease of use and perceived usefulness to the attitudes of individuals in using a system [30,59]. This discussion thus proposes the hypotheses below: Fig. 1 summarizes the conceptual model

Hypothesis 8: Perceived usefulness is linked positively to attitude towards social media usage.
Hypothesis 9: Perceived ease-of-use is linked positively to attitude towards social media usage.
Hypothesis 10: Perceived enjoyment is linked positively to attitude towards social media usage.

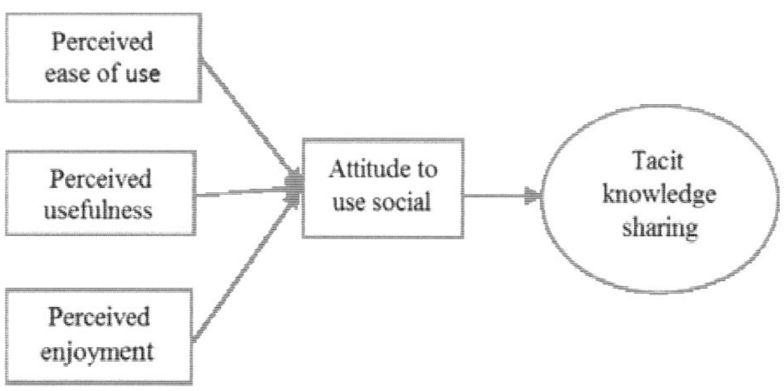

Figure 1. Conceptual tacit knowledge sharing model

4. Methodology

This paper is organized in the form of a general review, which includes study designs that are used to identify the best evidence to tackle the research objective. Scientific publications were selected and grouped according to their respective aims and contents, which eventually comes down to three categories namely: (i) TAM literature reviews and extension of TAM, (ii) social media, and (iii) tacit knowledge sharing. Scientific publications concerning TAM are identified based on a structured approach namely: (i) search on particular keyword(s) in leading databases, (ii) selection of publications with the corresponding criteria, and (iii) a quick scan of the titles, abstracts, and full text of the selected publications to determine their relevance; and (iv) reading and analyzing the full text of the selected publications. The resources were purposely examined to determine the analytical gap and to understand the social media applications for the disbursement of knowledge sharing.

The search was contained within English-language studies that have been published. Comprehensive electronic search had used the ISI Web of Knowledge and Google Scholar. Although scattered across journals and conferences, the studies appeared more frequently in journals such as Computers in Human Behavior, Computers & Education, Communications of the Association for Information Systems, Decision Support Systems, Expert Systems with Applications, Government Information Quarterly, Information & Management, International Journal of Electronic Government Research, and MIS Quarterly. The analysis of the articles is based on a series of characteristics consisting of the types of correlations between the model's constructs, external variables, limitations of studies, and methodological details.

5. Discussion and conclusion

Based on the findings and the hypotheses developed in the previous sections, a new conceptual framework (See Fig. 1) has been proposed with regards to the correlation between the TAM initiatives and the behavior of tacit knowledge sharing in social media. The proposed model shows that these features create opportunities for an effective flow of tacit knowledge in a social media setting. The model facilitates a better understanding of the phenomenon of tacit knowledge sharing via social web initiatives as well as provides a new avenue for discussion in this area. The model provides an important link between the requirements of tacit knowledge sharing and the contribution of social media in complying with those requirements, which has not been investigated in any literature. There are several aspects to the model which can be criticized. For example, the inclusiveness of the model is questionable because the process of tacit knowledge sharing is quite complicated and is influenced by various surrounding conditions. Furthermore, empirical evidences for justifying the hypothesized relationships are rare and require testing. Despite the shortcomings above, the model gives way for new theoretical grounds which brings us closer to fully exploring the applications of online social tools for the sharing of tacit knowledge. This study suggested a research model based on the technology acceptance theory.

Current literature on social media provides studies that were conducted from a technology acceptance model perspective, without the provision of a holistic understanding of user intention specifically with regards to tacit knowledge sharing in an online setting. Hence, the literature analysis in this study indicated a direct link between perceived ease-of-use, perceived usefulness and perceived enjoyment with regards to the sharing of tacit knowledge. Furthermore, the mediation role of Attitude in the use of social media was hypothesized in the relations between perceived ease-of-use, perceived usefulness and perceived enjoyment. The findings of this study bring several noteworthy implications. This study contributes to the body of knowledge in the field of IT acceptance and particularly the sharing of tacit knowledge in social media. For the first time, this study hypothesizes the mediation effect of Attitude on the TAM construct and tacit knowledge sharing. This study could be regarded as a working paper. Subsequent empirical studies could be conducted to verify the findings of this study. The correlations in conceptual frameworks need to be validated and tested in various social media contexts.

References

[1] Nonaka I A 1994 Dynamic theory of organizational knowledge creation. *Organ. Sci* , vol. 5, pp. 14-37.
[2] Elizabeth A, Aaron S 2001 The role of tacit and explicit knowledge in the workplace. *Journal of knowledge Management*, vol. 5, pp. 311-321.
[3] Callahan S Want to manage tacit knowledge? Communities of practice offer a versatile solution. White Paper Anecdote.[online] available: http://www. anecdote.com.
[4] Steininger K, Ruckel D, Dannerer E and Roithmayr F 2010 Healthcare knowledge transfer through a web 2.0 portal: An Austrian approach. *Int J Healthc Tech Manag*, vol 11, pp. 13-30.
[5] Hsia T L, Lin L M, Wu J H and Tsai H T 2006 A framework for designing nursing knowledge management systems. Interdisciplinary *Journal of Information, Knowledge, and Management*, vol. 1, pp. 13-23.
[6]. Zheng Y, Li L and Zheng F 2010 In Social media support for knowledge management, *International Conference on Management and Service Science*,; pp 1-4.
[7]. Panahi J, Watson S and Partridge H 2016 Social media and physicians: Exploring the benefits and challenges. *Health Inform. J* vol. 22, pp. 99-112.
[8] Stewart S A and Abidi S S R 2012 Applying social network analysis to understand the knowledge sharing behaviour of practitioners in a clinical online discussion forum. *J Med Internet Res*, vol. 14, pp. 170-179.

[9] Amidi A, Jusoh Y Y., Abdullah R H, Jabar M A, Khalefa M S A 2015 In: An overview on leveraging social media technology for uncovering tacit knowledge sharing in an organizational context, Software Engineering Conference (MySEC), 2015 9th Malaysian,; IEEE: pp 266-271.
[10] Vuori V and Okkonen J V 2012 Knowledge sharing motivational factors of using an intra-organizational social media platform. *Journal of Knowledge Management*, vol. 16, pp. 592-603.
[11] Matschke C, Moskaliuk J, Bokhorst F, Schummer T and Cress U. C 2014 Motivational factors of information exchange in social information spaces. Comput.*Human Behav, vol.* 36, pp. 549-558.
[12] Nielsen P and Razmita L P 2014 Motivation and knowledge sharing through social media within danish organizations. In Creating value for all through it, Springer; pp 197-213.
[13] Yuan D, Lin Z and Zhuo R D 2016 What drives consumer knowledge sharing in online travel communities?: Personal attributes or e-service factors? *Comput. Human Behav*, vol. 63, pp. 68-74.
[14] Panahi S, Watson J and Partridge H 2015 Information encountering on social media and tacit knowledge sharing. *J. Inf. Sci*, 0165551515598883
[15] Levy M 2013 Stairways to heaven: Implementing social media in organizations. *Journal of Knowledge Management*, 17, 741-754.
[16] Smith G 2010 Social software building blocks. Retrieved November, vol. 5.
[17] Freeman B and Chapman S B 2007 Is “youtube” telling or selling you something? Tobacco content on the youtube video-sharing website. *Tob. Control*, vol. 16, pp. 207-210.
[18] David I, Poissant L and Rochette A I 2012 Clinicians’ expectations of web 2.0 as a mechanism for knowledge transfer of stroke best practices. *J Med Internet Res*, vol. 14,
[19] Gormley S 2012 Understanding participation in knowledge-sharing in virtual communities of practice on the hseland elearning portal [thesis]/by sandra gormley. University of Dublin, Trinity College,.
[20] Podgorski D 2010 The use of tacit knowledge in occupational safety and health management systems. *Int J Occup Saf Ergon* , vol. 16, 283-310.
[21] Abidi S S R, Cheah Y N and Curran J S 2005 A knowledge creation info-structure to acquire and crystallize the tacit knowledge of health-care experts. *Inf Technol Biomed*, IEEE Transactions on, vol. 9, pp. 193-204.
[22] Hau Y S., Kim B, Lee H and Kim Y. G 2013 The effects of individual motivations and social capital on employees’ tacit and explicit knowledge sharing intentions. *Int. J. Inf. Tech.*, vol. 33, pp. 356-366,
[23] Luoma S and Okkonen J S 2009 In Capturing competence–using wiki for transferring tacit knowledge, Proceedings of the European Conference on Intellectual Capital in Haarlem, The Netherlands,; pp 329-336.
[24] Antheunis M L, Tates K and Nieboer T E 2013 Patients’ and health professionals’ use of social media in health care: Motives, barriers and expectations. Patient education and counseling, vol. 92, pp. 426-431.
[25] Sarringhaus M M 2011The great divide: Social media's role in bridging healthcare's generational shift. *Int J Healthc Manag.,* vol. 56, pp. 235-244,
[26] Chung L H 2001 In The role of management in knowledge transfer, Third Asian Pacific Interdisciplinary Research in Accounting Conference,; Citeseer: pp 15-17.
[27] Hildreth P M and Kimble C 2002 The duality of knowledge. Information research, vol. 8.
[28] Sharratt M and Usoro A M 2003 Understanding knowledge-sharing in online communities of practice. *EJKM*, 1, 187-196.
[29] Mun Y Y, Jackson J D, Park J S and Probst J C 2006 Understanding information technology acceptance by individual professionals: Toward an integrative view. *Information & Management*, vol. 43, pp. 350-363,
[30] Kim S 2012 Factors affecting the use of social software: Tam perspectives. *Electronic Library*, vol. 30, pp. 690-706.

[31] Davis F D 1989 Perceived usefulness, perceived ease of use, and user acceptance of information technology. MIS quarterly, pp. 319-340.
[32] Soo K 2006 Why workers share knowledge: A case study. Unpublished doctoral dissertation. Indiana University.
[33] Choi J H, Lev B and Kim H L 2014 In Exploring determinants of knowledge sharing in a social network of practice, Proceedings of the Seventh *International Conference on Management Science and Engineering Management*,; Springer: pp 39-51.
[34] Gefen D and Straub D W 2000 The relative importance of perceived ease of use in is adoption: A study of e-commerce adoption. *J. Assoc. Inf. Syst*, vol. 1, pp. 8.
[35] Hanson C, West J, Neiger B, Thacheray R, Barnes M and McIntyre E C 2011 Use and acceptance of social media among health educators. *A J Health Educ.*, vol. 42, pp. 197-204,
[36] Hsu C L and Lin J C 2008 Acceptance of blog usage: The roles of technology acceptance, social influence and knowledge sharing motivation. *Information & Management*, vol. 45, pp. 65-74,
[37] Albusaidi K and Olfman L 2014 In Knowledge workers' attitude toward inter-organizational knowledge sharing system in the education sector, Proceedings of the 19 th UKAIS *Conference on Information Systems*, Oxford, England.
[38] A. Bilgihan, A. Barreda, F. Okumus, 2016 Consumer perception of knowledge-sharing in travel-related online social networks. *Tour Manag*, vol. 52, pp. 287-296,
[39] McGowan B S, Wasko M, Vartabedian B S, Miller R S, Freiherr D D and Abdolrasulnia M 2012 Understanding the factors that influence the adoption and meaningful use of social media by physicians to share medical information. *J Med Internet Res*, 14,
[40] Li H and Liu Y H 2014 Understanding post-adoption behaviors of e-service users in the context of online travel services. *Information & Management*, vol. 51, pp. 1043-1052,
[41] Davis F D and Bagozzi R P 1992 Extrinsic and intrinsic motivation to use computers in the workplace1. J. *Appl. Soc. Psychol*, vol. 22, pp. 1111-1132.
[42] Hung H T and Yuen S C 2010 Educational use of social networking technology in higher education. *Teaching in higher education*, vol. 15, pp. 703-714.
[43] Hong S J, Thong J Y, J Moon Y, Tam K Y 2008 Understanding the behavior of mobile data services consumers. *Inf Syst Front.,* vol. 10, pp. 431-445.
[44] Sun H, Zhang P 2006 Causal relationships between perceived enjoyment and perceived ease of use: An alternative approach. *J. Assoc. Inf. Syst*, vol. 7, pp. 24.
[45] Zhou T 2011 Understanding online community user participation: A social influence perspective. *Internet Res* , vol. 21, pp. 67-81.
[46] Wasko M M 2005 Why should i share? Examining social capital and knowledge contribution in electronic networks of practice. MIS quarterly, pp. 35-57.
[47] Yu T K, Lu L C and Liu T F 2010 Exploring factors that influence knowledge sharing behavior via weblogs. *Comput Human Behav*, vol. 26, pp. 32-41,
[48] Papadopoulos T, Stamati T and Nopparuch P 2013 Exploring the determinants of knowledge sharing via employee weblogs. *Int. J. Manag. Rev.*, vol. 33, pp. 133-146.
[49] Kankanhalli A, Tan B C and Wei K K A 2005 Contributing knowledge to electronic knowledge repositories: An empirical investigation. MIS quarterly, pp. 113-143.
[50] Kang M and Schuett M A 2013 Determinants of sharing travel experiences in social media. J *Travel Tour Mark*, vol. 30, pp. 93-107.
[51] Nonaka I and Takeuchi H I 1995 The knowledge-creating company: How japanese companies create the dynamics of innovation. Oxford university press.
[52] Ma Q and Liu L 2004 The technology acceptance model: A meta-analysis of empirical findings. *J. Org. & End User Comp* (JOEUC), vol. 16, pp. 59-72.
[53] Yousefzai S Y, Foxall G R and Pallister J G 2007 Technology acceptance: A meta-analysis of the tam: Part 1. *Journal of Modelling in Management*, vol. 2, pp. 251-280.
[54] Islam M A, Agrawal N. K and Ikeda M 2014 Library adoption of knowledge management using web 2.0 a new paradigm for libraries. IFLA journal, 40, 317-330.

[55] Lin K Y and Lu H P 2011 Why people use social networking sites: An empirical study integrating network externalities and motivation theory. *Comput. Human Behav*, vol. 27, pp. 1152-1161.
[56] Lee W, Xiong L and Hu C 2012 The effect of facebook users' arousal and valence on intention to go to the festival: Applying an extension of the technology acceptance model. *Int J Hosp Manag.*, vol. 31, pp. 819-827.
[57]. Rosen L D, Carrier L M and Cheever N. A 2013 Facebook and texting made me do it: Media-induced task-switching while studying. *Comput. Human Behav*, vol. 29, pp. 948-958.
[58]. Venkatesh V and Davis F D 2000 A theoretical extension of the technology acceptance model: Four longitudinal field studies. *Manag. Sci*, vol. 46, pp. 186-204.
[59]. Hsu C L and Lu H P 2004 Why do people play on-line games? An extended tam with social influences and flow experience. *Information & management*, vol. 41, pp. 853-868.
[60] Nor A M F, Azmi J, Rusli A, Masrah A and Azmi M 2015 Preliminary Study of Knowledge Management (KM) Practices in Malaysian Automotive Industry. *Journal of Computer Science & Computational Mathematics*, vol. 5, pp. 13-20.

The Impacts of Demographic Variables on Technological and Contextual Challenges of E-learning Implementation

Hanan Aldowah[1, 2], **Samar Ghazal**[1, 2], **Irfan Naufal Umar** [1], **Balakrishnan Muniandy** [1]

[1]Universiti Sains Malaysia (USM), Centre for Instructional Technology and Multimedia, (CITM), Penang, Malaysia
[2]Centre for Information System and Technology (CIST), Hodiedah University, Hodiedah, Yemen

hanan_aldwoah@yahoo.com

Abstract. Information technology has achieved robust growth which has made it possible for learning to occur quickly. The rapid development of information, communication and technologies (ICT) has initiated an unparalleled transformation in universities all over the world. This development of technology and learning is offering new techniques to represent knowledge, new practices, and new global communities of learners. As a result, today's economic and social changes force universities to try to find new learning approaches and systems. E-learning seems to be an appropriate approach in this aspect. However, the implementation of e-learning systems in universities is not an easy task because of some challenges related to context, technology, and other challenges. This paper studied the impacts of demographic data and reported the critical points for the decision makers to consider when planning and implementing e-learning in universities. A quantitative approach was used to study the effects of technological and contextual challenges on e-learning implementation in which a questionnaire was used for the data collection. According to the findings of the study, the most important challenges of the implementation of e-learning are related either to organizational (Contextual) and technological (technical) issues. The demographic variables have been found to play a direct and indirect role with the technological and contextual challenges of implementing e-learning. This paper showed that there are some significant differences in the two challenges faced by instructors in terms of the demographic variables. The result revealed that some significant differences exist between demographic variables and the two challenges of e-learning in terms of gender, age, teaching experience, ICT experience and e-learning experience. However, there is no significant difference in terms of e-learning experience. The obtained data, from such study, can provide information about what academic institutions can do before implementing e-learning to reduce and overcome the challenges in implementing e-learning in universities. So, university administrators interested in implementing e-learning should recognize the challenges that their instructors are facing and to provide the necessary policy and support to help overcome these challenges.

1. Introduction

Information technology has achieved robust growth which has made it possible for learning to occur quickly. The rapid development of information, communication and technologies (ICT) has initiated an unparalleled transformation in universities all over the world. E-learning is a result of the integration of technology in education, which enables the creation of innovative research and delivery mechanisms which is one of the features of modern learning [1]. Innovative thinking about acquiring necessary skills and knowledge as well as the ability to manage efficiently within the available resources are required to meet new issues and challenges [2].

In the last ten years, the use of e-learning has been increasing rapidly and has become an important system offered by most colleges and universities all over the world. This comes as the consequence of the innovative transfer of knowledge and learning which has been influenced by the advent of the Internet and other ICT [2]. E-learning is currently the common term used to describe the various uses of information and communications technologies to enhance learning and teaching. It takes the form of online courses and training, by which the courses and training are delivered via the Internet [3, 4].

Universities have been motivated to adopt e-learning to transform traditional learning environments and create more efficient and attractive learning experiences [5]. Successful implementation of e-learning requires an understanding of the issues that promote and support the effective use of the new technology [6-8]. The implementation of e-learning in universities should be as part of educational reform process. For effective application of e-learning, a great deal of attention needs to be given its implementation [9]. When universities promote e-learning use, they need to understand their instructors' and students' attitudes towards its use. Instructors' attitudes are considered as a major predictor of the use of new technologies in instructional settings [10].

While e-Learning provides several benefits to educational settings which enhance the quality of education and develop the learning environments, there remain many challenges and issues which hinder the exploration and utilization of its opportunities [11-13].

Technological and contextual challenges are considered two of the main challenges that face the implementation of e-learning in universities. Technological challenges are defined as the challenges surrounding issues of familiarity with new technology and technical problems encountered by instructors and other users [14] such as, Internet bandwidth, and computer facilities. While contextual challenges refer to the issues and problems that are related to the context of e-learning includes the context of the delivering organization (university setting) and the context of society in which e-learning take place including culture, rules and regulations [8]. It requires adjustments not only from the organization's side but also from the instructors' and student's sides [15]. A lack of easy access to these technologies as well as a lack of skills necessary to use available computer and communication technology has hindered instructors' use of e-learning [16].

[17] and [18] have pointed out that the technological challenges of the e-learning framework involving issues of technology infrastructure in e-learning environments. A lack of easy access to necessary computer equipment, Internet connectivity and other technologies are among the critical problems in the implementation of e-learning [16, 19].

It also refers to the quality and reliability of the connectivity [20] as well as the bandwidth will affect the users' ability to access the full range of the content needed [2]. Along with challenges such as electricity, skills, and training of staff on the use of new technology [8].

Moreover, the cost of the technologies needed in setting up the e-learning system is considered a limitation to the successful implementation of e-learning [8, 20, 21]. According to [15], technology is dominant in e-learning, yet it is also expensive. This makes the initial costs of buying computers, especially in the early implementation stages as well as the ongoing costs of upgrading systems, very high. Therefore, this factor is frequently mentioned in the literature as one of the most significant barriers to e-learning.

In addition, there are contextual challenges (organizational and societal/cultural challenges) which include: the effect of instructors perception and attitude toward e-learning, the effect of beliefs of decision makers on e-learning, lack of understanding the new roles of instructor and student, and the need for proper rules and regulation for e-learning implementation[16, 22].

Unfortunately, universities are still away from getting the full benefit and advantage of e-learning because of these technological and contextual challenges [4, 23]. These challenges are fundamental and critical that need to be identified and addressed for the successful implementation of e-learning [4, 17, 24].

The literature indicates that these challenges are associated with the instructors' cultural background and attitudes towards e-learning; the high cost of technological infrastructure; the often prohibitive cost of educational technologies and the lack of local expertise in curriculum development for e-learning [8, 25-27] which are asserted by previous studies [2, 20, 28] .

Many researchers have argued that examining individual differences or aspects is a significant research area in e-learning [29]. In addition, the impacts of demographics on e-learning implementation and the use of e-learning in higher education are well documented [30, 31]. The developers of e-learning environments are constantly advised by the researchers to address demographic differences regarding the use of e-learning. The differences are based on the individual-characteristics of gender, age, ICT experience, and e-learning experience [1, 30, 31].

2. Research Objective

The main objective of this study is:

- To examine if there are significant differences in the technological and contextual challenges faced by the instructors in terms of the demographic variables of:
  i. Gender
  ii. Age group
  iii. Teaching Experience
  iv. ICT Experience.
  v. E-learning Experience

3. Research Question

The research question of this study is:

RQ 1 Are there significant differences in the technological and contextual challenges faced by the instructors in terms of the demographic variables of:

i. Gender
ii. Age group
iii. Teaching Experience
iv. ICT Experience
v. E-learning Experience.

The following hypothesis was generated by the researcher from the research question:

H_0: *There are no significant differences in the technological and contextual challenges faced by the instructors in terms of the demographic variables of: Gender; Age group; teaching experience; ICT experience; e-learning experience.*

4. Methodology

The objective of this research paper is to examine if there are significant differences between the technological and contextual challenges of implementing e-learning and the demographic profiles of instructors. The population of the study comprises 107 instructors of Hodiedah University, Yemen (2014/2015). These **107** instructors are from all the sixteen faculties of the university and the sample included males / females, and junior / senior instructors. The instrument consists of 53 items and the participants are asked to rate their agreement of each item of the survey on a 4-point Likert Scale 1-4.

Validity was first conducted to measure if the questionnaire structure and content accurately achieves the purpose for which it was designed. To achieve this purpose, research experts and academicians group assessed the instrument and the homogeneity of the variables before it was used.

Thirty questionnaires were sent to experts and academicians group of faculty members in different schools and centres in USM (Universiti science Malaysia) in the field of e-learning, Computer Science and Instructional Technology to ensure content validity. The instrument also reviewed by the experts in the e-learning centre at Hodiedah University who with wide experience and keen interest in e-learning.

The research instrument's reliability was assessed using Cronbach's alpha (α). Alpha values for the two constructs exceeded the 0.70 lower limit of acceptability [32]. For technological challenges construct is 0.837, and for contextual challenges is 0.858.

In this study, inferential statistics were used to identify any significant differences between the technological challenges and contextual challenges of implementing e-learning and the demographic variables of the participants. In this study, there are five demographic variables involving:

Gender, Age, Teaching experience, ICT experience, E-learning experience (instructors who have already had teaching experience with e-learning courses.

The technological challenges of implementing e-learning include three factors as follows:

1. Access to computer and Internet
2. Cost of technology
3. Power Supply.

The contextual challenges of implementing e-learning included two main constructs, involving organizational and social/cultural. The organizational construct (Org_CH) consists of three factors:

1. knowledge management,
2. economy and funding,
3. training of instructors.

The second construct is societal/cultural (So_CH). It consists of five factors:

1. roles of instructors and students,
2. participants attitude on e-learning & IT,
3. roles and regulation.

In this paper, the significant differences between the technological challenges and contextual challenges of implementing e-learning with the demographic variables were analyzed by using Analysis of variance (ANOVA) and T-test. The one-way ANOVA and T-test were employed to identify the effects of demographic variables on the technological challenges of implementing e-learning.

The selection of the appropriate statistics was based on the distribution of the dependent variable as well as on the number of groups being compared. One-way ANOVA measures the mean of one or more groups based upon independent demographic variables. On the other hand, t-test is usually employed to evaluate the difference in the means between two groups. In the present research, one-way ANOVA is used for age, field of teaching (faculty), teaching experience, whilst t-test is used for gender, ICT experience, and e-learning experience. A **p** value of less than **0.05** (**$p<0.05$**) indicated when the differences are significant.

4.1. Effect of Gender on E-learning Challenges

H_{01}: *There are no significant differences in the technological and contextual challenges faced by the instructors in term of gender.*

For this null hypothesis, the t-test was used to determine the significant differences among the mean scores of the independent groups (Male and Female). The results are shown in Table 1

Table 1. The result of significant differences between the challenges and gender

Challenge	Gender	N	Mean	Std. Deviation	t	df	*Sig.
Technological Challenges	Male	67	3.54	0.532	2.022	105	0.046
	Female	40	3.34	0.451			
Contextual Challenges	Male	67	3.59	0.388	2.449	105	0.016
	Female	40	3.40	0.415			
95% Confidence Interval of the difference			*sig (p<005)				

The result indicates a significant difference in technological challenges among male and female (**t**: 2.02, **p:** 0.046) with the male instructors perceived that e-learning has more technology challenges than female instructors (male: 3.54, female: 3.34).

In term contextual challenges, the result also indicates a significant difference among male and female (**t**: 2.44, **p:** 0.016) with the male instructors perceived that e-learning has more contextual challenges than female instructors (male: 3.59, female: 3.40). Both results imply to males have given more importance to these challenges than females Thus, the null hypothesis was rejected.

4.2. Effect of Age on E-learning Challenges

H_{02}: *There are no significant differences in the technological and contextual challenges faced by the instructors in terms of age.*

In this study, five age groups have been investigated as shown in Table 2.

Table2. Age of respondents

Age group	Frequency	Percent
25-35	42	39.3
36-46	40	37.4
47-57	23	21.5
More than 57	2	1.9
Total	**107**	**100.0**

For this null hypothesis, the analysis of variance (ANOVA) was used. The results are shown in Table 3.

IOP Conf. Series: Journal of Physics: Conf. Series **892** (2017) 012013 doi:10.1088/1742-6596/892/1/012013

Table 3. The result of significant differences between e-learning challenges and age

Challenges	Age group	Sum of Squares	df	Mean Square	F	Sig. *
Technological Challenges	Between Groups	0.853	3	0.284	1.09	0.356
	Within Groups	26.828	103	0.260		
	Total	27.681	106			
Contextual Challenges	Between Groups	1.539	3	0.513	3.290	0.024
	Within Groups	16.060	103	0.156		
	Total	17.599	106			
95% Confidence Interval of the difference					***sig (p<005)**	

Table 3 indicates there is no significant difference in technological challenges among the different age groups (f: 1.09, p: 0.356). However, the result indicates that there is a significant difference in contextual challenges among the age groups (**f**: 3.29, **p**: 0.024). Therefore, the null hypothesis was partially rejected. Table 4 shows the mean scores for all age groups.

All the age groups in terms of technological challenges and contextual challenges recorded high scores as shown in Table 4.

Table 4. Description of mean scores of e-learning challenges with age

Categories	Age Group	N	Mean	Std. Deviation
Contextual Challenges	25-35	42	3.44	0.362
	36-46	40	3.67	0.375
	47-57	23	3.42	0.484
	More than 57	2	3.22	0.157
	Total	107	3.52	0.407
Technological Challenges	25-35	42	3.37	0.512
	36-46	40	3.51	0.542
	47-57	23	3.58	0.454
	More than 57	2	3.25	0.000
	Total	107	3.47	0.511

4.3. Effect of Teaching Experience on E-learning Challenges

H_{03}: *There is no significant difference in the technological and contextual challenges faced by the instructors in term of teaching experience.*

In this study, six age groups have been investigated as shown in Table 5.

Table 5. Teaching experience of respondents

Year group	Frequency	Percent
1-5	30	28.0
6-10	35	32.7
11-15	26	24.3
16-20	12	11.2
21-25	2	1.9
26-30	2	1.9
Total	**107**	**100**

In order to test this null hypothesis, a one-way ANOVA was performed between-groups. The results are shown in Table 6.

Table 6. The result of significant differences between e-learning challenges and Teaching Experience

Challenges	Teaching experience	Sum of Squares	df	Mean Square	F	*Sig.
Technological Challenges	Between Groups	1.481	5	0.296	1.142	0.343
	Within Groups	26.200	101	0.259		
	Total	27.681	106			
Contextual Challenges	Between Groups	3.732	5	0.746	5.436	0.000
	Within Groups	13.867	101	0.137		
	Total	**17.599**	**106**			
95% Confidence Interval of the difference		***sig (p<005)**				

Results in Table 6 indicate that there is no significant difference in the technological challenges with the different teaching experience groups (**f**: 1.14, **p**: 0.343). However, the result indicates that there is a significant difference in contextual challenges among the teaching experience groups (**f**: 5.43, **p**: 0.000). Therefore, the null hypothesis was partially rejected. Table 7 shows the mean scores for all teaching experience groups.

Finding in Table 7 indicates that instructors with difference number of years of experiences faced a high level of technological challenges where the mean scores ranged from 3.00 to 3.61. The highest mean score recorded by those who have teaching experience between " 11-15 " and the lowest mean score is 3.00 for the those who have teaching experience between " 21 -25 " which still considers high challenges faced by instructors in the university.

In terms of contextual challenges, the mean scores ranged from **2.83** to **3.70**. In addition, the result indicated that those who have teaching experience between " 11-15" years of teaching recorded the highest mean score **3.70,** while those who have teaching experience between " **21 -25** " recorded the lowest mean score **2.83.**

Table 7. Description of mean scores of e-learning challenges with Teaching Experience

Challenges	Years	n	Mean	Std. Deviation
Technological challenges	1-5	30	3.35	0.429
	6-10	35	3.48	0.576
	11-15	26	3.61	0.496
	16-20	12	3.52	0.548
	21-25	2	3.00	0.354
	26-30	2	3.25	0.000
	Total	**107**	**3.47**	**0.511**
Contextual Challenges	1-5	30	3.32	0.341
	6-10	35	3.64	0.362
	11-15	26	3.70	0.371
	16-20	12	3.44	0.434
	21-25	2	2.83	0.707
	26-30	2	3.22	0.157
	Total	**107**	**3.52**	**0.408**

4.4. *Effect of E-learning Experience on E-learning Challenges*

H_{05}: *There is no significant difference in the technological and contextual challenges faced by the instructors in terms of e-learning experience.*

In order to test this null hypothesis, a t-test was performed to determine the significant differences among the mean scores of the independent groups. The results are shown in Table 8.

Table 8. The result of significant differences between e-learning challenges and e-learning experience

Challenges	E-learning Experience	N	Mean	Std. Deviation	df	t	sig
Technological Challenges	Yes	43	3.52	0.497			
	No	64	3.43	0.521	105	0.967	0.336
	Total	**107**	**3.47**	**0.511**			
Contextual Challenges	Yes	43	3.55	0.440	105	0.701	0.485
	No	64	3.50	0.386			
	Total	**107**	**3.52**	**0.407**			

95% Confidence Interval of the difference *sig **(p<005)**

Findings in Table 8 indicate that there is no significant difference in terms of technological challenges between those with e-learning experience and those without e-learning experience (**t**: 0.967, **p**: 0.336). Similarity, there is no significant difference was found in terms of contextual challenges between those with e-learning experience and those without e-learning experience (**t**: 0.0.70, **p**: 0.485). Therefore, the null hypothesis was completely supported.

Table 8 also shows the mean scores for the two groups (instructors with e-learning experience and those without any e-learning experience).

5. Discussion of the Effect of Instructors Profiles on the Technological and Contextual Challenges of E-learning Implementation

The demographic profiles of instructors have been found to play an important role with the challenges of implementing e-learning. Much research is dedicated to investigating the effect between individual differences and e-learning implementation [23, 28]. These differences emerge from individual characteristics such as gender, age, teaching experience, previous training and experiences. Thus, this section discusses the results of the differences between the technological and contextual challenges of implementing e-learning based on selected demographics variables. Some demographics of instructors that have been selected for this study included gender, age, teaching experience, ICT experience and e-learning experience.

The findings of this research paper revealed that there is a significant difference between male and female instructors views regarding the technological and contextual challenges that effect the implementation of e-learning. In the current research, male instructors faced more challenges than instructors female. This may be due to male instructors taking these challenges into consideration when deciding whether or not to implement e-learning into their teaching. A significant gender variation was found supporting prior research [33, 34].

In term of age, the findings revealed that there are significant differences between the instructors' views from different age groups regarding the contextual challenges of implementing e-learning. These results imply that the age of respondents has an effect on contextual challenges, which include perception and attitude of instructors toward the implementation of e-learning. However, the results also showed that the age of respondents has no significant difference with technological challenges of e-learning implementation. The effect of age on e-learning challenges was found in previous research by [35, 36]. This is true with the majority of participants between 25 to 46 years old. These age groups have less challenge regarding technological issues.

With regards to teaching experience, the result revealed that there is significant differences for contextual challenges with different teaching experience groups. This outcome is in agreement with previous research [1, 37]. These findings suggested that teaching experience affects the implementation of e-learning in terms of context. The results also revealed that there is no significant difference with technological challenges of e-learning implementation. This is because majority of the participants are from less experience groups. Thus, this finding implies that participants with less teaching experience face less technological challenges. This result is consistent to the U.S National center for Education Statistics (2000) which reported that instructors with less experience in teaching faced less challenges related to technology and they are more likely to use and implement technology in their teaching than instructors with more experience in teaching.

With regards to ICT experience, all but one respondent reported having ICT experiences and high level of technological challenges. Although ICT experience plays an important role in the implementation of e-learning, the technical skills that are necessary to implement e-learning are different. It can be concluded from this finding that instructors need intensive course-based technology training, which takes them beyond the basic computer skills to effectively embed technology into the learning process. [2, 31].

In term of e-learning experience, the instructors sampled recorded no significant difference between the e-learning experience and the technological and contextual challenges of implementing e-learning. In other words, the findings revealed that instructors with e-learning experience and those without any e-learning experience hold the same views in relation to the technological and contextual challenges of implementing e-learning. This may be due to the fact that instructors with e-learning experience faced more challenges in terms of technological challenges than those who without any experience. This implies that previous experience in e-learning does not effect in the implementation of e-learning. To put differently, whether there is an e-learning experience or no experience the challenges still exist. This is consistent with prior research [1].The findings of this research revealed that there was no significant differences between e-learning challenges with different levels of e-learning experience.

It can be concluded from this finding that the skills that are necessary in e-learning systems differ when the system is in the implementation phase or when it is in the normal operation. Integrating e-learning process into the university or any organization is not an easy task and requires specialist with various skills to be in the implementation team. An e-learning project should bring together specialist with pedagogical, technical, creative skills. This outcome is in agreement with previous research [38].

6. Conclusion

This paper investigated the impacts of demographic data and reported the critical points for the decision makers to consider when planning and implementing e-learning in universities. According to the findings of the study, the most important challenges of the implementation of e-learning are related either to organizational (Contextual) and technological (technical) issues. Administrators should consider the needs of instructors when implementing e-learning in any university. For instance, they will need training programs on computers and e-learning. Moreover, instructors need to have a technical support while they interact in the e-learning environment. Taking into account the negative impact of all the technological and contextual challenges that might affect the implementation of e-learning. In addition, instructors' experiences with using e-learning technologies were low. Therefore, instructors should be trained in order to keep them current with the emerging educational technologies. This can be accomplished by providing training courses about e-learning technology tools such as training programs on some important software and/or hardware for e-learning. In addition, providing workshops on e-learning for instructors is much needed to encourage them to be familiar with the modern teaching approaches.

References

[1] Taha, M., Investigating the success of E-learning in secondary schools: The case of the Kingdom of Bahrain. 2014, Brunel University.
[2] Qureshi, I.A., et al., *Challenges of implementing e-learning in a Pakistani university.* Knowledge Management & E-Learning: An International Journal (KM&EL), 2012. **4**(3): p. 310-324.
[3] Al-Shboul, M., *The level of e-Learning integration at The University of Jordan: Challenges and opportunities.* International Education Studies, 2013. **6**(4): p. p93.
[4] Yoloye, E.O., *New Technologies For Teaching and Learning: Challenges For Higher Learning Institutions In Developing Countries.* Information Communication Technology (ICT) Integration to Educational Curricula: A New Direction for Africa, 2015: p. 250.
[5] Abdelraheem, A.Y., *The Implementation of E Learning in the Arab Universities: Challenges and Opportunities.* Proceedings of DLI 2006, 2006: p. 145-155.
[6] Bhuasiri, W., et al., *Critical success factors for e-learning in developing countries: A comparative analysis between ICT experts and faculty.* Computers & Education, 2012. **58**(2): p. 843-855.

[7] Goi, C.L. and P.Y. Ng, *E-learning in Malaysia: Success factors in implementing e-learning program.* International Journal of Teaching and Learning in Higher Education, 2009. **20**(2): p. 237-246.
[8] Aldowah, H., S. Ghazal, and B. Muniandy, *Issues and Challenges of Using E-Learning in a Yemeni Public University.* Indian Journal of Science and Technology, 2015. **8**(32).
[9] Ilias, I.S.H.C., M.S.I.S. Salim, and S.F. Sulaiman, *Mobile Learning: A QoS study on Video Conferencing over WiFi/WiMAX network.* Journal of Computer Science & Computational Mathematics, 2014. **4**(2): p. 4.
[10] Gulbahar, Y. and F. Kalelioglu, *Competencies for e-Instructors: How to Qualify and Guarantee Sustainability.* Contemporary Educational Technology, 2015. **6**(2).
[11] Alkharang, M.M. and G. Ghinea, *E-learning in Higher Educational Institutions in Kuwait: Experiences and Challenges.* E-learning, 2013. **4**(4).
[12] Kwofie, B. and A. Henten. *The advantages and challenges of e-learning implementation: The story of a developing nation.* in *WCES-2011 3rd World Conference on Education Sciences, Bahcesehir University, Istabul, Turkey*. 2011.
[13] Mapuva, J., *Confronting challenges to e-learning in higher education institutions.* International Journal of Education and Development Using ICT, 2009. **5**(3): p. 101-114.
[14] Olutola, A.T. and O.O. Olatoye, *Challenges of E-Learning Technologies in Nigerian University Education.* Journal of Educational and Social Research, 2015. **5**(1): p. 301.
[15] Ali, G.E. and R. Magalhaes, *Barriers to implementing e-learning: a Kuwaiti case study.* International journal of training and development, 2008. **12**(1): p. 36-53.
[16] Wang, S.-C., B. Cowie, and A. Jones. *Challenges of e-learning for university instructors in Taiwan.* in *The 16th International Conference on Computers in Education (ICCE 2008). Taipei, Taiwan*. 2008. Citeseer.
[17] Tarus, J.K., D. Gichoya, and A. Muumbo, *Challenges of implementing e-learning in Kenya: A case of Kenyan public universities.* The International Review of Research in Open and Distributed Learning, 2015. **16**(1).
[18] Khan, B., *Learning features in an open, flexible and distributed environment.* AACE Journal, 2005. **13**(2): p. 137-153.
[19] Aleman-Meza, B., et al., *An ontological approach to the document access problem of insider threat*, in *Intelligence and Security Informatics*. 2005, Springer. p. 486-491.
[20] Kamba, M., *Problems, challenges and benefits of implementing e-learning in Nigerian universities: An empirical study.* International Journal of Emerging Technologies in Learning (iJET), 2009. **4**(1).
[21] Andersson, A., *Seven major challenges for e-learning in developing countries: Case study eBIT, Sri Lanka.* International Journal of Education and Development using ICT, 2008. **4**(3).
[22] Rhema, A., I. Miliszewska, and E. Sztendur. *Attitudes towards e-learning and satisfaction with technology among engineering students and instructors in Libya*. in *Proceedings of the Informing Science and Information Technology Education Conference*. 2013.
[23] Surry, D.W. and D.C. Ensminger, *Facilitating the Use of Web Based Learning by Higher Education Faculty.* Online Submission, 2006.
[24] Mahmud, K. and K. Gope. *Challenges of implementing e-learning for higher education in least developed countries: a case study on Bangladesh*. in *Information and Multimedia Technology, 2009. ICIMT'09. International Conference on*. 2009. IEEE.
[25] Muthanna, A. and A.C. Karaman, *Higher education challenges in Yemen: Discourses on English teacher education.* International Journal of Educational Development, 2014. **37**: p. 40-47.
[26] Aljaaidi, K.S.Y., *GIRLS'EDUCATIONAL CRISIS SOLVING THROUGH THE ADOPTION OF E-LEARNING SYSTEM: THE CASE OF HADHRAMOUT UNIVERSITY.* 2009.

[27] Thabet, T.S.A. and N. Kalyankar, *The Effect of E-Learning Approach on Students' Achievement in Fraction Math Course Level 5 at Yemens Public Primary School.* Global Journal of Computer Science and Technology, 2014. **14**(2).
[28] Al-Harbi, K.A.-S., *E-Learning in the Saudi tertiary education: Potential and challenges.* Applied Computing and Informatics, 2011. **9**(1): p. 31-46.
[29] Agarwal, R. and J. Prasad, *Are individual differences germane to the acceptance of new information technologies?* Decision sciences, 1999. **30**(2): p. 361-391.
[30] Abouchedid, K. and G.M. Eid, *E-learning challenges in the Arab world: Revelations from a case study profile.* Quality Assurance in Education, 2004. **12**(1): p. 15-27.
[31] Nawaz, A., *The Challenges & Opportunities of eLearning.* 2011.
[32] Hair, J., et al., *Multivariate data analysis (6 ed.). Englewood Cliffs, NJ: Pearson Prentice Hall.* 2006.
[33] Ong, C.-S. and J.-Y. Lai, *Gender differences in perceptions and relationships among dominants of e-learning acceptance.* Computers in Human Behavior, 2006. **22**(5): p. 816-829.
[34] Spotts, T.H., M.A. Bowman, and C. Mertz, *Gender and use of instructional technologies: A study of university faculty.* Higher Education, 1997. **34**(4): p. 421-436.
[35] Adelabu, O.A., E. Adu, and S. Adjogri. *The Availability and Utilization of E-Learning Infrastructures for Teaching and Learning.* in *World Conference on Educational Multimedia, Hypermedia and Telecommunications*. 2014.
[36] 36. Osika, E., R. Johnson, and R. Butea, *Factors influencing faculty use of technology in online instruction: A case study.* Online Journal of Distance Learning Administration, 2009. **12**(1).
[37] Islam, M.A., et al., *Effect of demographic factors on e-learning effectiveness in a higher learning institution in Malaysia.* International Education Studies, 2011. **4**(1): p. p112.
[38] Cech, P. and V. Bures. *E-learning implementation at university.* in *Proceedings of the 3rd European Conference on e-learning, Paris, France*. 2004.

The Awareness and Challenges of Cloud Computing Adoption on Tertiary Education in Malaysia

Nor Hazreeni Hamzah, Maziah Mahmud, Shamsunarnie Mohamed Zukri, Wan Fairos Wan Yaacob and Jusoh Yacob

Faculty of Computer and Mathematical Sciences, UiTM Cawangan Kelantan, Malaysia

reeni683@ kelantan.uitm.edu.my

Abstract: This preliminary study aims to investigate the awareness of the adoption of cloud computing among the academicians in tertiary education in Malaysia. Besides, this study also want to explore the possible challenges faced by the academician while adopting this new technology. The pilot study was done on 40 lecturers in Universiti Teknologi MARA Kampus Kota Bharu (UiTMKB) by using self administered questionnaire. The results found that almost half (40 percent) were not aware on the existing of cloud computing in teaching and learning (T&L) process. The challenges confronting the adoption of cloud computing are data insecurity, data insecurity, unsolicited advertisement, lock-in, reluctance to eliminate staff positions, privacy concerns, reliability challenge, regulatory compliance concerns/user control and institutional culture/resistance to change in technology. This possible challenges can be factorized in two major factors which were security and dependency factor and user control and mentality factor.
Keywords: cloud computing, awareness, challenges, factor analysis.

1. Introduction

Computer technology has become an essential tool for educators to ensure that teaching and learning (T&L) process becomes more interesting and effective. The using of cloud computing will enhance T&L process which is differ to traditional techniques. Cloud computing is the interactive way for educators to collaborate with students along T&L process. Nowadays, social writing platforms appear to be logistically useful tools for a variety of campus needs, from student group learning to faculty department work to staff collaborations. By using computer technology in the classroom, will allow educators to explore new and exciting methods of instruction and learning beyond lecturing. There are a lot of useful and interactive website that can enhance classroom lessons.

A study conducted by [1] showed a comprehensive comparison between e learning system before and after moving onto Cloud to see the contribution of e learning standards with the Cloud standards. The result support the moving of e learning on to Cloud Computing environment where Cloud Computing is used as e learning solutions to delivery of computing as a service rather than a product.

The impact of cloud computing is that the adoption of cloud computing has a significant impact on cost effectiveness where it can reduce development team cost, technical support team cost, burden of daily backup management and cost of overall project expenditure [1, 2, 6, 3]. Furthermore, the adoption of cloud computing revealed that it will enhanced availability, low environmental impact, reduced IT complexities, mobility, scalability, increased operability and reduced investment in physical asset [2]. [7] analyzed three different perspectives of scheduling algorithms for energy efficiency which is heuristic, game theory and learning process strategy. As a result, they found out that heuristic strategy is a common method for scheduling tasks while managing the energy consumed. Energy management structure for cloud computing in order to develop the optimal energy efficient scheduling are identified through similarities, differences and energy models. They believed that the solution to green cloud computing is between energy consumption and system performance.

Comprehensive literature review on the cloud computing research had been conducted by [4] for 236 scholarly journals articles between 2009 and 2014. The main purpose of this study is to investigate the awareness and challenges of cloud computing adoption among the academicians in tertiary education in Malaysia.

2. Cloud Computing

Cloud computing allowed users to have access to data, applications and services anywhere. It also allows users to use application without installation of personal files. Nowadays, industries and educational areas are adopting the services of cloud computing because of the following reasons:

- Low Cost - with the cloud, the users will pay for applications only when needed and many applications are included free of charge.
- Scalability.
- Ease of access - Quite simply, cloud computing is easy to get up and running. Instead of having to download and/or install software yourself, in the cloud it is all done for you.
- No time boundation - This allows for on demand analysis of study material instantly.
- High performance - For using the cloud computing the performance should be enhanced. Cloud computing implies a level of dynamic, flexible resources, sharing and allocation of assets.
- Cloud computing has four key concepts which are, service oriented architecture/thin clients, delivery models for cloud computing, virtualization and deployment strategies.

Model of cloud computing are as follows:

- Infrastructure as a Service (IaaS) - focused on hardware and IT infrastructure management which includes storage and processing capabilities.
- Platform as a Service (PaaS) - Concentrates on middleware which provides development tools and hosting options for cloud clients to manage their own applications. PaaS gives end users control over application design, but not control over physical infrastructure.
- Software as a Service (SaaS) - includes software applications which support business related process like customer relationship management, supply chain management or enterprise resource planning. SaaS does not give end users control over application designs.

The illustration of the model of Cloud Computing is presented in Figure 1 [5].

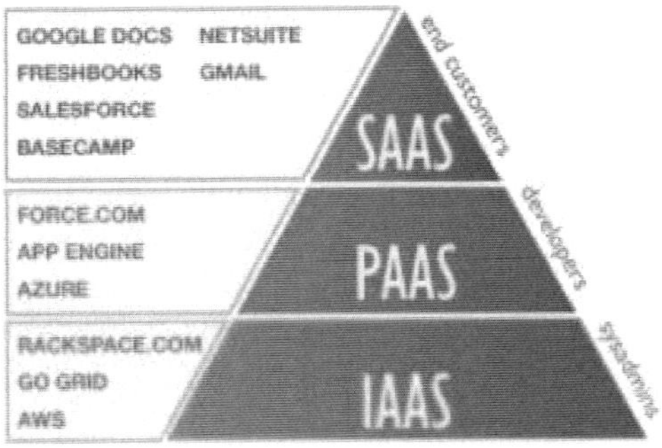

Figure 1. Illustration of the Layers for the Services Oriented Architecture.

3. Materials and Methods

The pilot study was conducted among UiTMKBs lecturers, from January to March, 2017. A validated administered questionnaire was used in this study. The questionnaire was adapted from [2]. The questionnaire consisted of three sections which were demographics profiles, awareness and challenges section. A total of 40 lecturers completed the questionnaire.

The reliability of questionnaire was tested by using the reliability analysis (Cronbach alpha). The Cronbach alpha for section challenges was 0.763 which indicated the instrument was reliable for data collection.

4. Statistical Analysis

Prevalence was calculated for awareness of the adoption of cloud computing among lecturers in UiTMKB. Factor analysis used to describe covariance relationship among challenges item in term of a few underlying but unobservable factors. A two sided hypothesis was used in analysis with $p \leq 0.05$ was considered statistically significant. All the data were analyzed using IBM SPSS version 20.

5. Results

There was 24 lecturers were aware on the adoption of cloud computing in T&L process (prevalence =60%), whereas 15 were female and the rest were male. Those who adopt IAAS, SAAS and PAAS were 20%, 55% and 10% respectively.

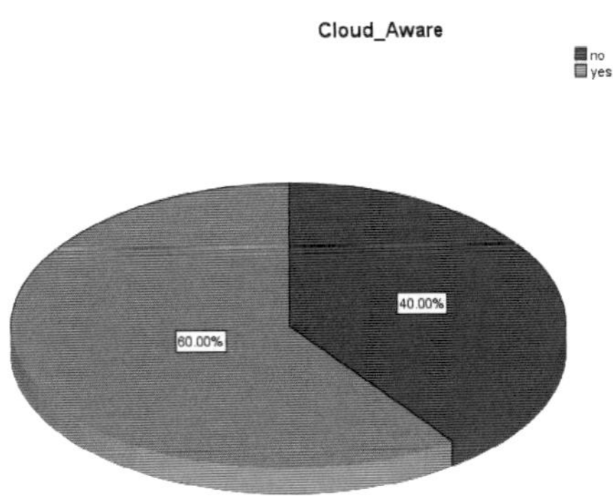

Figure 2. Cloud Computing Awareness

Initially the factorability of the eight items in section challenge was examine. Several well recognize criteria for the factorability of the correlation were used. Firstly, all of eight items correlated since all of the correlation values >0 .3, suggesting reasonable factorability. Secondly, the Kaiser-Meyer-Olkin measure of sampling adequacy was .505, above the recommended value of .5, and Bartlett's test of sphericity was significant (χ^2 (28) = 87.900, p < .05). Finally, the communalities were all above .3 (see Table 1), further confirming that each item shared some common variance with other items. Given these overall indicators, factor analysis was conducted with all 8 items.

Principle components analysis was used because the primary purpose was to identify and compute composite coping scores for the factors underlying the challenges. The initial eigen values showed that the first factor explained 38.6% of the variance and the second factor 30.4% of the variance. The remaining six factors explained 31.0% of the variance. The factor solutions were examined, using varimax rotations of the factor loading matrix. The two factor solution, which explained 69.0% of the variance, was preferred because of its previous theoretical support, the 'leveling off' of eigen values on the scree plot after two factors (see Figure 3), and the insufficient number of primary loadings and difficulty of interpreting the subsequent factors.

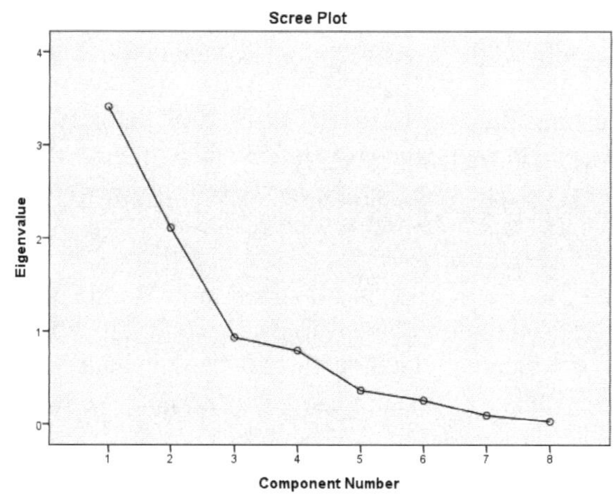

Figure 3. Component Number

Out of eight challenges items, based on the rotated component matrix (see Table 1), these possible challenges can be factorized into two groups. The items in first factor were Data insecurity, Unsolicited Advertisement, Reluctance to eliminate staff positions, Privacy concerns, and Reliability challenge. While the items belong to second factor was Lock-in, Regulatory compliance concerns/ User control and Institutional culture/resistance to change in technology.

Table 1. Extraction Method (Principal Component Analysis)

	Communalities		Rotated Component	
	Initial	Extraction	1	2
Data insecurity	1	0.764	0.665	-0.566
Unsolicited Advertisement	1	0.679	0.814	0.127
Lock-in	1	0.545	-0.001	0.738
Reluctance to eliminate staff positions	1	0.391	0.592	-0.202
Privacy concerns	1	0.903	0.886	0.343
Reliability challenge	1	0.786	0.811	0.358
Regulatory compliance concerns/User control	1	0.722	0.434	0.73
Institutional culture/resistance to change in technology	1	0.73	0.048	0.853

6. Other recommendations

The prevalence of awareness of the adoption of cloud computing among lecturers in UiTMKB was 60%, which was considered higher than the previous study done among public universities [2]. Based on the finding, most of the academicians in UiTMKB were comfort with SAAS in adopting cloud computing in their T&L process. This study also suggests that the challenges faced by academician while adopting this technology can be separated into two major themes. First, the challenges can be named as security and dependency factor which consisted five items and the rest items was named as user control and mentality factor.

Acknowledgement

A special gratitude to our group members in stimulating suggestions, encouragement and coordinate on this preliminary project. Furthermore, we would like to acknowledge with much appreciation to UiTM especially on the opportunity of having ARAS grant. We would like to extend our sincere thanks to all respondents who participated in this mini project.

References

[1] F. F. Ahmed. Comparative Analysis for Cloud Based e-learning. In *Procedia Computer Science*, 2015.

[2] O. C. Akin, F. T. Matthew, and D. Comfort. The Impact and Challenges of Cloud Computing Adoption on Public Universities in Southwestern Nigeria. *IJACSA International Journal of Advanced Computer Science and Applications*, 5(8), 2014.

[3] M. Al-Zoube. E-Learning on the Cloud. *International Arab Journal of e-Technology*, 1(2):58-64, 2009.

[4] M. Bayramusta and V. A. Nasir. A fad or future of IT?: A comprehensive literature review on the cloud computing research. *International Journal of Information Management*, 36(4):635-644, 2016.

[5] A. Fern´andez, D. Peralta, F. Herrera, and J. M. Ben´ıtez. An overview of e-learning in cloud computing. In *Advances in Intelligent Systems and Computing*, 2012.

[6] N. M. Rao, C. Sasidhar, and V. Satyendra Kumar. Cloud Computing Through Mobile-Learning. *International Journal of Advanced Computer Science and Applications*, 1(6), 2010.

[7] Norhashimi Bin Mohd Nor, Masnida Binti Hussin, Hasan Bin Selamat, 2014, Energy Management for Cloud Computing: A Survey from Scheduling Perspective of Heuristic, Game Theory and Learning Strategy. *Journal of Computer Science & Computational Mathematics,* 4(3):42-48, 2014.

Similarity Measure for Molecular Structure: A Brief Review

S A Bero, A K Muda, Y H Choo, N A Muda and S F Pratama

Computational Intelligence and Technologies (CIT) Research Group, Center of Advanced Computing and Technologies, Faculty of Information and Communication Technology, Universiti Teknikal Malaysia Melaka Hang Tuah Jaya, 76100 Durian Tunggal, Melaka, Malaysia

sitiasmah.bero@ gmail.com

Abstract. Similarity or distance measures have been used widely to calculate the similarity or dissimilarity between two samples of dataset. Cheminformatics is known as the domain that dealing with chemical information and both similarity and distance coefficient have been an important role for matching, searching and classification of chemical information. There are various types of similarity/distance coefficient used in molecular structure similarity searching. Generally, the calculation using similarity/distance coefficient is focusing more on 2-dimensional (2D) rather than 3-dimensional (3D) structure. In this paper, the popular similarity/distance coefficients for molecular structure will be reviewed together with the review on 3D molecular structure.

1. Introduction

The Cambridge English Dictionary defines similar to be "looking or being almost, but not exactly, the same" and the comparison of objects to determine their levels of similarity lies at the heart of many academic disciplines. The "Similarity" also said as the resemblance, likeness, sameness and an agreement. The concept of similarity searching was being applied to numerous types of application and domain, such as in pattern recognition, chemistry and cheminformatics. The high demand for drug discovery research recently makes the molecular structure searching is one of the prominent research areas in the cheminformatics.

The earliest study of similarity searching on molecular structure started in mid- eighties that was conducted by [1] Raymond et al.,that proposed an algorithm for computing the simple type of substructure similarity known as atom pairs. A year later, the similarity studies are continued by [2, 3] until now.

The similarity searching of molecular structure is closely related to the *Similar Property Principle,* whereby, molecules that are structurally similar are likely to have similar properties [4]. This relationship lead to the range of cheminformatics techniques as it had established a discipline that has applications in many areas of chemistry and computational molecular science [5, 6] such as molecular diversity analysis, clustering, QSAR/QSPR methods and also similarity searching [4].

In order to conduct the similarity searching, there are three important components of similarity measures that need to be considered [7]. First, the structure representation (used to characterize molecules), second, the weighting scheme (used to determine the relative importance value of the representation), and lastly is similarity coefficient. The similarity coefficient plays an important role in the context of similarity search, as it is used to quantify the degree of resemblance between the two structural representations [4].

The molecular representation is always being the first thing to consider before conducting any similarity searching in molecular structure. There are various types of descriptors reported [8-10] that had been used in the studies of molecular similarity. According to [4], the descriptors can be divided into three classes: 1D, 2D and 3D representations.

A molecular structure consists of a chemical formula that indicates the numbers and types of atoms in a molecule. Whereby, each atom is linked together by a bond to form a complete molecule structure. A molecular structure is an important element in the area of similarity searching as it is bringing the description of type, arrangement, position and direction of the bonds linking atom within a molecule [11]. Besides that, the molecular structure is the fundamental unit in studying the chemical information of molecule. Whereby, variety types of chemical compounds will uniquely identify by their molecular structure [12].

A molecular structure can be expressed in either 1D, 2D or 3D [13]. Many researchers have worked with 2D molecular structure rather than 3D structure, for example, in the study of molecular structure similarity searching [14-17]. This is due to the fact that 2D structure is simpler, easy to work on, and not as complicated as 3D. A molecular structure becoming a main component in the drug discovery whereby that chemical information extracted from the structure is very valuable for the chemist.

Furthermore, a molecular structure been an important aspect in drug discovery, because many kind of drugs can be identified based on their consistent structure of its molecular structure. Although there are many research has been done on the 2D molecular structure, however less attention is given to 3D structure similarity searching. Therefore, the study of 3D structure similarity searching is still relevant and open for exploration as the 3D structure will capture all aspects of molecular structure that 2D does not, which are the size of molecular structure and the shape of molecular structure.

This paper presents on some of the popular similarity/distance measurer used for similarity searching of molecular structure. At the same time, this paper also reviews on 3D molecular structure. The paper will be organized as follows: Section 2 introduces the 3D molecular structure, followed by an analysis and a timeline of the similarity measurer in Section 3. Finally, the conclusions are given in Section 4.

2. 3D Molecule Structure

A chemical structure can be represented either in 2D or 3D, as shown in figures 1 and 2 respectively. Before the emergence of 3D structure, the 1D and 2D has been widely used in previous research on chemical structure; especially in similarity search of chemical structure. Those structures become a priority previously because both 1D and 2D structure allows for ease of developing an algorithm [18] and easy to work on. Later, this section will briefly discuss on the characteristic of 3D molecular structures in the context of similarity searching.

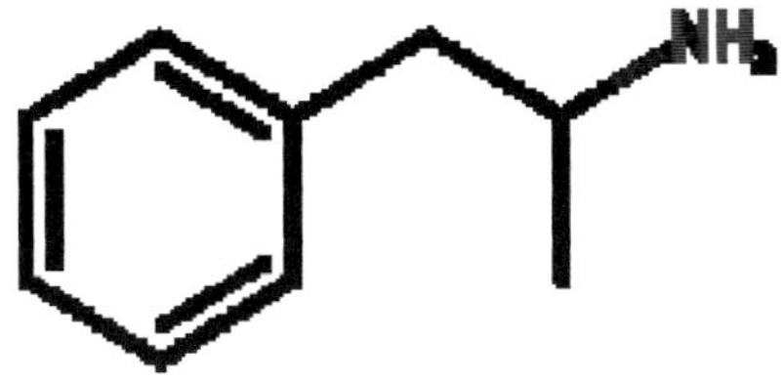

Figure 1. An example of 2D molecule structure of amphetamine

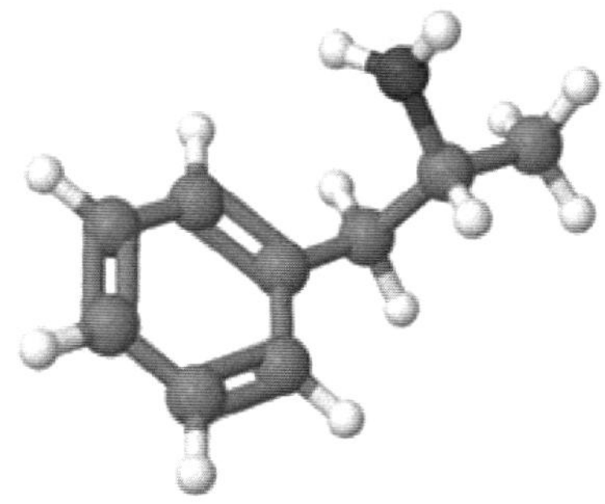

Figure 2. An example of 3D molecule structure of amphetamine

2.1. Characteristics of 3D Chemical structure

According to [19], the molecular similarity is a complex structure that can only be described with reference to the immediate use for which it is intended. Similarity encompasses the bonding patterns, atomic positions, conformation, shape and spatial disposition of molecular properties [19]. This is simply saying that, to undergo a 3D similarity chemical structure, it is needed to consider the whole properties of 3D chemical structure itself.

In the context of 3D similarity searching, it requires the conformation properties and the specification of an entire target structure (the whole structure) rather than partial structure. Thus, the 3D method-based similarity measurement was introduced and gained more attention recently because of their potential to overcome the key limitation of 1D and 2D methods [13].

The 3D structures usually have three different angles of view, which is *x*-, *y*-, and *z*- axis. Meanwhile, the 2D structure only has two angles of view; *x*- and *y*-axis. With the one extra angle of view in 3D structure, it will provide an important information needed and better conformation for the similarity search process. Another characteristic of 3D structure is, it is better in capturing all aspects of the molecular structure (the size of molecular structure and the shape of molecular structure). In order to measure how similar is the chemical structure is, a similarity measure is needed.

3. Similarity and Distance

A similarity or distance measure can be defined as a tool to quantify the similarity or dissimilarity between object or known as a function that can compute the degree of similarity or dissimilarity between a pair of object. Numerous similarity measures and distance measures have been used widely in various fields of the area. For example, in social network [20], text similarity [21], document similarity [22], triangle inequality [23], biology [24, 25], image retrieval [26-28], fingerprint [29, 30], handwritten character [31, 32], and chemistry [14-17, 33].

From mathematical point of view, *distance* is used to define how far between two objects are [34]. Meanwhile, from the metric point of view, the distance also known as *dissimilarity*, whereby the concept is still the same; to find the distance between two objects. In terms of distance coefficients, the distance is used to quantify the degree of difference between two objects. The smaller value of distance, the larger the degree of similarity and vice versa [35].

Figure 3 below shows the similarity measure and distance measure that is commonly used in 2D binary feature vector that was arranged in historical order [36]. Those measures are divided into three different classes; one class of distance based, and two classes of similarity measure (non-correlation based and correlation based).

Based on figure 3, the majority of the similarity measure is grouped under the Correlation based. Jaccard, Russell & Rao, Dice & Sorenson, Tanimoto, Roger, Sokal & Michener, Hamann, Faith and Gower & Legendre are grouped under Non- Correlation coefficient in [36]. Meanwhile, this Non-Correlation coefficient is grouped as an Association coefficient in [35] and there are 16 of them,

IOP Conf. Series: Journal of Physics: Conf. Series **892** (2017) 012015 doi:10.1088/1742-6596/892/1/012015

including Jaccard/Tanimoto, Dice, Russell/Rao, Sokal/Sneath (1)-(3), Kulczynski (1)-(2), Simple matching, Hamann, Rogers/Tanimoto, Baroni-Urbani/Buser, Ochiai/Cosine, Forbes, Fossum and Simpson.

3.1. Past Studies of 3D Similarity Searching

The 3D molecular structure similarity search is believed to overcome the limitation of 1D and 2D structure on the aspect of retrieving important information which cannot be retrieved on both 1D and 2D structures. Hence, the 3D molecular structure similarity search becomes a concern nowadays, especially in the drug discovery.

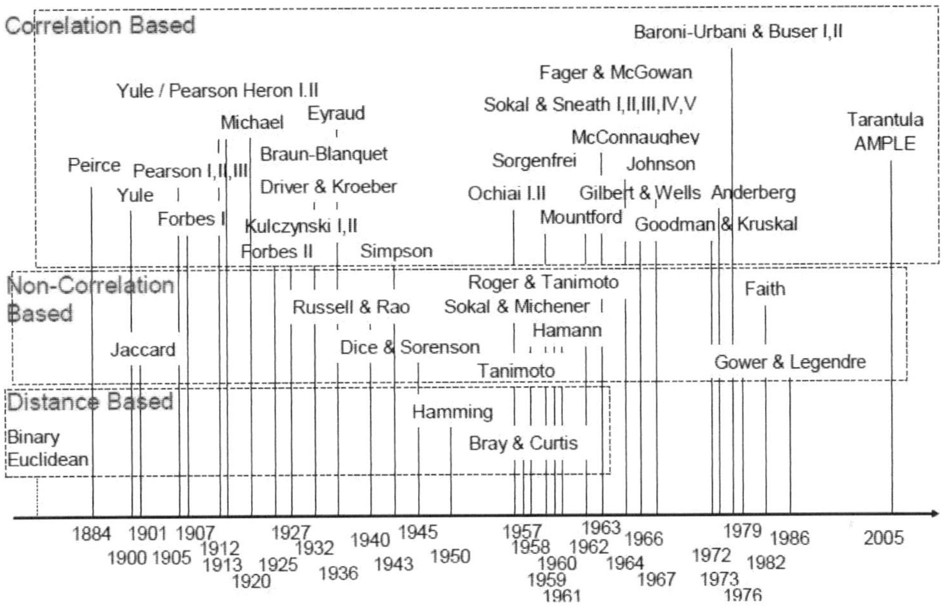

Figure 3. The chronological of binary similarity and distance measures by year [36]

The 3D molecular structure similarity searching has begun since more than 2 decades ago and a numerous method and approach has been introduced. For example, in the early 1980s, [37] had done a matching measure between two molecular structures based on the density function, and followed by [38], which used an application namely SPERM (Superpositioning by PERMutations), to search a 3D molecular structure of two biological molecules namely netropsin and daunomycin in 30 000 compound databases.

Meanwhile, [39] have proposed four different algorithms of pattern matching to determine the presence or absence of an atom and associated interatomic distance in 3D chemical structure and the results show that, out of four algorithms proposed, the Ullman algorithm is more efficient on small molecular structure searching compared to the others.

The study of similarity searching for 3D molecular structure represented by interatomic distance matrices is continued in the year of 1990s. The study done by [40] had compared between atom mapping and clique detection method for the calculation of 3D structural similarity. From the experiment, the atom mapping method is more efficient in small 3D molecular similarity searching, compared to the Clique detection is well measured for 3D macromolecular similarity searching. Later, the study in 3D molecular structure similarity searching keeps rising year by year as reported by [14, 41-44].

Nowadays, the similarity searching of 3D molecular structure had rapidly developed with various applications and methods which are reviewed in [13]. Commonly, the 3D representation was measured based on their shape similarity [45], which compared to 2D similarity, most similarity calculation is based on 2D fingerprints (0 and 1) [4]. The similarity/distance coefficient commonly used in 2D, also being used in 3D. For example, Manhattan distance, Tanimoto coefficient, and cosine coefficient [13].

3.2. Analysis of Similarity Measurement for Molecular Structure

According to [7], a similarity searching will be effective if, and only if, an appropriate similarity measure is employed. Of the similarity measure listed in the previous study, a similarity measure under an association coefficient is commonly used in similarity searching of molecular structure, for example, the Tanimoto coefficient [14]. The association coefficient is specifically designed for use with binary data, which used 0 and 1 to indicate the absence and the presence of molecular structure [33].

Table 1. Common similarity/distance measure in chemical informatics

Similarity/Distance Coefficient	Formula	
Tanimoto Coefficient	$Tani = \frac{c}{a+b-c}$	(1)
Dice Coefficient	$Dice = \frac{2a}{2a+b+c}$	(2)
Cosine Coefficient	$Cos = \frac{a}{\sqrt{(a+b)(a+c)}}$	(3)
Euclidean Distance	$ED(a,b) = \sqrt{\sum_{i=1}^{n}(b_i - a_i)^2}$	(4)
Hamming Distance	$HM(a,b) = a+b-2c$	(5)

The nature of association coefficient usually working with a binary data and often normalized between the range of zero (no similarity at all) and unity (identical set of descriptors) [33, 35]. Besides the binary data, the association coefficient also can be used with non-binary data, in which the different formulation and range of value may apply [35]. Tanimoto coefficient, Dice coefficient, Cosine coefficient, Euclidean distance and Hamming distance also being known as the most similarity measures used in similarity searching of molecular structure studies.

Equation (1) refers to the Tanimoto coefficient (also known as the Jaccard coefficient) and it said as the extension from the Jaccard coefficient [16]. The Tanimoto coefficient is defined as the sum of the common feature occur in both structures (c) over the total feature that occur only in a and b, minus with the common feature, c.

Most of the previous research by research group at the University of Sheffield used the Tanimoto coefficient as the similarity measure in lots of their experiment on similarity searching of the chemical structure as presented in [2-4, 7, 9, 14, 33]. The reason why the Tanimoto coefficient becomes a priority in the context of similarity of molecular structure was presented in [40]. The Tanimoto coefficient was preferred based on two reasons; firstly, on the basis of a subjective evaluation of the similarity search ranking, and secondly, referred to its calculation that does not involve any square root, making it faster in calculation [14].

The ease of implementation and speed is the main reason why a Tanimoto coefficient becomes the most popular similarity measure in the context of chemical informatics and computational medicinal chemistry [46]. Besides that, the other similarity measure such as Dice coefficient, Cosine coefficient,

Euclidean distance and Hamming distance also being used in similarity searching of molecular structure but not rapidly as the Tanimoto coefficient. The Dice coefficient in equation (2) also referred as the Sorensen, Czekannowski, Hodgkin-Richard or Morisita [34]. The Dice coefficient measure the similarity directly like Tanimoto coefficient [10] making it ease in implementation and fast in the calculation.

Equation (3) indicates the Cosine coefficient that is used to measure the similarity between two vectors by measuring the cosine of the angle between the data objects and then, the similarity is measured as the angle between them [47]. Compared to Tanimoto and Dice coefficient, the Cosine coefficient is often used to measure the similarity between sparse data (like document or binary data). For example, text mining and measure the cohesion within clusters in the field of data mining [47]. In the context of chemical similarity, the Cosine coefficient allows the calculation of the average similarity between all pairs of compound in the dataset very rapidly which is impossible when using Tanimoto coefficient [14].

Meanwhile, the Euclidean distance and Hamming distance are distance coefficient that been used to quantify the degree of dissimilarity between two molecules. Both of these coefficients effectively in identifying the common absence of the feature as evidence of similarity [10], and [14] stated that, the Hamming distance is equivalent to the squared Euclidean distance for binary data. Meanwhile, the Euclidean distance is very useful for measuring the similarity between dense data such as time series or two dimensional points [47]. It is reported as the Euclidean distance remains the most popular measure of distances either for continuous data or count data [16].

4. Conclusion

This paper reviewed the common similarity/distance measurer in molecular structure. The Tanimoto coefficient is found the most widely similarity measures used in similarity searching of chemical informatics fields compared to the other similarity measure. Besides that, the majority of the similarity and distance measurer is commonly used for binary calculation that is referred more on 2D chemical structure. Meanwhile, the similarity searching of 3D chemical structure must dealing with their conformational properties and shape representation in order to find the similarity. Some researchers believed that the similarity and distance measurer can be used for non-binary data (especially on the association coefficient), but the modification to the range of the value is need. In conclusion, the studies of similarity searching in chemical informatics offer the advantages to the chemist and also in the field of drug discovery itself.

Acknowledgments

This work is funded by Universiti Teknikal Malaysia Melaka (UTeM) through the PJP High Impact Research Grant [PJP/2016/FTMK/HI3/S01473].

References

[1] Carhart R E, Smith D H, and Venkataraghavan R 1985 *Atom pairs as molecular features in structure-activity studies: definition and applications* (*J. Chem. Inf. Comput. Sci.* vol 25) chapter 2 pp 64–73
[2] Willett P, Winterman V, and Bawden D 1986 *Implementation of nearest-neighbor searching in an online chemical structure search system* (*J. Chem. Inf. Model* vol 26) pp 36–41
[3] Willett P and Wintermann V 1986 *A Comparison of Some Measures for the Determination of Inter-Molecular Structural Similarity Measures of Inter-Molecular Structural Similarity* (*Quant. Struct. Relationships* vol 25) pp 18–25
[4] Willett P 2014 *The calculation of molecular structural similarity: Principles and practice* (*Mol.Inform.* vol 33) chapter 6–7 pp 403–413
[5] Arulmozhi, Chemoinformatics V 1907 *A Quick Review* pp 416-419
[6] Begam B F and Kumar J S 2012 *A Study on Cheminformatics and its Applications on Modern Drug Discovery (Procedia Engineering 38*(Icmoc)) pp 1264–1275

[7] Willett P 2009 *Similarity methods in chemoinformatics* (*Annu. Rev. Inf. Sci. Technol.* vol 43), pp 1–117
[8] Todeschini R and Consonni V 2002 *Handbook of Molecular Descriptors* Wiley-VCH, Weinheim
[9] Warr W E 2011*WIREs Computational Molecular Sciences* vol 1 pp 557
[10] Leach A R and Gillet V J 2007 *An introduction to Chemoinfornatics*
[11] Molecular structure. (n.d.) *Jonas: Mosby's Dictionary of Complementary and Alternative Medicine*. (2005), http://medicaldictionary.thefreedictionary.com/molecular+structure
[12] Chemical structure, https://en.wiktionary.org/wiki/chemical_structure
[13] Shin W H, Zhu X, Bures M, and Kihara D 2015 *Three-Dimensional Compound Comparison Methods and Their Application in Drug Discovery* (*Molecules* vol 20) chapter 7 pp 12841–12862
[14] Willett P, Barnard J M and Downs G M 1998 *Chemical similarity searching* (*Journal of chemical information and computer sciences* vol *38) chapter* 6 pp 983-996
[15] Hagadone T R 1992 *Molecular substructure similarity searching: efficient retrieval in two dimensional structure databases* (*Journal of chemical information and computer sciences* vol *32)* chapter 5 pp 515-521
[16] Willett P 2006 *Similarity-Based Virtual Screening Using 2D Fingerprints (Discovery* vol 1*)* (New York: Universities of Leeds , Sheffield and York) pp 1046–1053
[17] Flower D R 1998 *On the Properties of Bit String-Based Measures of Chemical Similarity* (*Journal of Chemical Information and Modeling* vol *38)* chapter 3 pp 379–386
[18] Shin W H, Zhu X, Bures M G and Kihara D 2015 *Three-dimensional compound comparison methods and their application in drug discovery (Molecules* vol *20)* chapter 7 pp 12841-12862
[19] Dean P M 2012 *Molecular Similarity in Drug Design* vol 65 chapter 3
[20] Rawashdeh A and Ralescu A L 2015 *Similarity Measure for Social Networks-A Brief Survey* In *MAICS* pp 153-159
[21] Mihalcea R, Corley C and Strapparava C 2006 *Corpus-based and knowledge-based measures of text semantic similarity* (In *AAAI* vol 6) pp 775-780
[22] Huang A 2008 *Similarity measures for text document clustering* (In *Proceedings of the sixth new zealand computer science research student conference (NZCSRSC2008), Christchurch, New Zealand)* pp 49-56
[23] Lipkus A H 1999 *A proof of the triangle inequality for the Tanimoto distance (Journal of Mathematical Chemistry* vol *26)* chapter 1-3 pp 263-265
[24] Hubalek Z 1982 *Coefficients of Association and Similarity, Based on Binary (Presence Absence) Data: An Evaluation* (Biological Reviews vol 57) chapter 4 pp 669-689
[25] Gray K R, Aljabar P, Heckemann R A, Hammers A and Rueckert D 2013 *Alzheimer's Disease Neuroimaging Initiative.: Random forest-based similarity measures for multi-modal classification of Alzheimer's disease.*(*NeuroImage* vol *65)* pp 167-175
[26] Smith J R and Chang S F 1996 *Automated binary texture feature sets for image retrieval* (*International Conf. Accoust., Speech, Signal processing, Atlantic, GA*)
[27] Subrahmanyam M, Maheshwari R P and Balasubramanian R 2012 *Local maximum edge binary patterns: a new descriptor for image retrieval and object tracking* (*Signal Processing* vol *92)* chapter 6 pp 1467-1479
[28] Cerra D and Datcu M 2012 *A fast compression-based similarity measure with applications to content-based image retrieval (Journal of Visual Communication and Image Representation* vol *23)* chapter 2 pp 293-302
[29] Ghany K K A, Hassanien A E and Schaefer G 2014 *Similarity measures for fingerprint matching (*In *International Conference on Image Processing, Computer Vision, and Pattern Recognition, USA*) pp 21-24
[30] Bajusz D, Rácz A and Héberger K. 2015 *Why is Tanimoto index an appropriate choice for*

fingerprint-based similarity calculations? (Journal of cheminformatics vol *7*) chapter 1 pp 1
[31] Rodríguez-Serrano J A and Perronnin F 2012 *A model-based sequence similarity with application to handwritten word spotting* (*IEEE transactions on pattern analysis and machine intelligence* vol *34)* chapter 11 pp 2108-2120
[32] Wang Q F, Yin F and Liu C L 2012 *Handwritten Chinese text recognition by integrating multiple contexts (IEEE transactions on pattern analysis and machine intelligence* vol *34)* chapter 8 pp 1469-1481
[33] Willett P 2011 *Similarity searching using 2D structural fingerprints* (*Chemoinformatics and computational chemical biology*) pp 133-158
[34] Cha S 2007 *Comprehensive Survey on Distance / Similarity Measures between Probability Density Functions* (*Int. J. Math. Model. Methods Appl. Sci* vol 1) chapter 4 pp 300–307
[35] Screening H T 2002 *Grouping Of Coefficients For The Calculation Of Inter-Molecular Similarity And Dissimilarity Using 2D Fragment Bit- strings* vol 5(New York: Universities of Leeds , Sheffield and York) pp 155–166
[36] Seung-Seok C, Sung-Hyuk C and Tappert C 2010 *A Survey of Binary Similarity and Distance Measures* (*Journal of Systemics, Cybernetics & Informatics* vol. *8) chapter* 1 pp 43–48
[37] Carbó R, Leyda L and Arnau M 1980 *How similar is a molecule to another? An electron density measure of similarity between two molecular structures* (*Int. J. Quantum Chem.*, vol *17)* chapter 6 pp 1185–1189
[38] Van Geerestein V J, Perry N C, Grootenhuis P D J and Haasnoot C A G 1990 *3D Database searching on the basis of ligand shape using the sperm prototype method* (*Tetrahedron Computer Methodology* vol *3)* chapter 6 PART C pp 595–613
[39] T Brint A and Willett P 1987 *Pharmacophoric pattern matching in files of 3D chemical structures: comparison of geometric searching algorithms* (*Journal of Molecular Graphics* vol *5)* chapter 1 pp 49–56
[40] Artymiuk P J, Bath P A, Grindley H M, Pepperrell C A, Poirrette A R, Rice D W, Thorner D. A, Wild D J, Willett P, Bank W, Sheffield S, Allen F H, and Taylor R 1992 *Similarity Searching* pp 617–630
[41] Bath P A, Poirrette A R, Willett P, Sheffield S and Allen F H 1994 *Similarity Searching in Files* pp 141–147
[42] Rouvray D H 1994 *Similarity studies. 1. The necessity of analogies in the development of science*(*Journal of Chemical Information and Modeling* vol *34)* chapter 2 pp 446–452
[43] Thorner D A, Willett P, Wright P M and Taylor R 1997 *Similarity searching in files of three dimensional chemical structures: representation and searching of molecular electrostatic potentials using field-graphs* (*Journal of Computer-Aided Molecular Design* vol *11)* chapter 96 pp 163–174
[44] Hahn M *1997Three-Dimensional Shape-Based Searching of Conformationally Flexible Compounds* (*Journal of Chemical Information and Modeling* vol *37)* chapter 96 pp 80–86
[45] Willett P, Finn and Morris G M 2013 WIRES Comp. Mol. Sci. vol 3 pp 226
[46] Maggiora G M, Vogt M, Stumpfe D, and Bajorath J J 2013 *Molecular similarity in medicinal chemistry* (*J. Med. Chem.* vol 57) chapter 8 pp 3186–3204
[47] M. A. Rabbih, S. M. A. Mamoun, Ezzat T. M. Selim, 2012, *Semiempirical Calculations of the Proton Affinities of Some Quinolone Derivatives* (Journal of Computer Science & Computational Mathematics vol. 2) pp. 53-55

Handling a Small Dataset Problem in Prediction Model by employ Artificial Data Generation Approach: A Review

Masitah Abdul Lateh[1], Azah Kamilah Muda[1], Zeratul Izzah Mohd Yusof[1], Noor Azilah Muda[1] and Mohd Sanusi Azmi[1]
[1]Computational Intelligence and Technologies Lab (CIT), Faculty of Information and Communication Technology, Universiti Teknikal Malaysia Melaka, 76100 Melaka, Malaysia

azah@utem.edu.my

Abstract. The emerging era of big data for past few years has led to large and complex data which needed faster and better decision making. However, the small dataset problems still arise in a certain area which causes analysis and decision are hard to make. In order to build a prediction model, a large sample is required as a training sample of the model. Small dataset is insufficient to produce an accurate prediction model. This paper will review an artificial data generation approach as one of the solution to solve the small dataset problem.

1. Introduction

In recent years, small dataset problem is widely discovered and getting more attention from researchers. There are many organizations that work with small datasets including manufacturing and medical areas. Manufacturing area usually dealing with small dataset problems in the early stage of manufacturing such as high cost of products while in the medical area the special medical record as an example, spinocerebellar ataxia, which is a kind of rare hereditary disease with very few records around the world.

There are many research works working on prediction model. To obtain a precise prediction model, a large sample set of data is required for the learning process. Otherwise, the model built is considered unreliable and the information gained is fragile [1]. For example, in the case of TFT-LCDs, their forecasting problem is related to a limited number of sample in the production line and a good prediction model could reduce the manufacturing cost. Thus, a solution is needed to handle a small dataset to build a precise prediction model and reduce a manufacturing costs. In Fig.1, a small dataset is one of the factors affecting inaccurate prediction model. So, if a small sample is used as training sample of a model, it might significantly affect the prediction uncertainty because of lack of information [2]. Consequently, the knowledge gained from small sample of data is considered unreliable and imprecise for learning system [3]. In addition, in the context of computational learning theory, small sample size gives a major effect to learning performances and is found as one of encountered in machine learning and data mining. This is because insufficient data size of training dataset is liable to poor performances of learning[4].

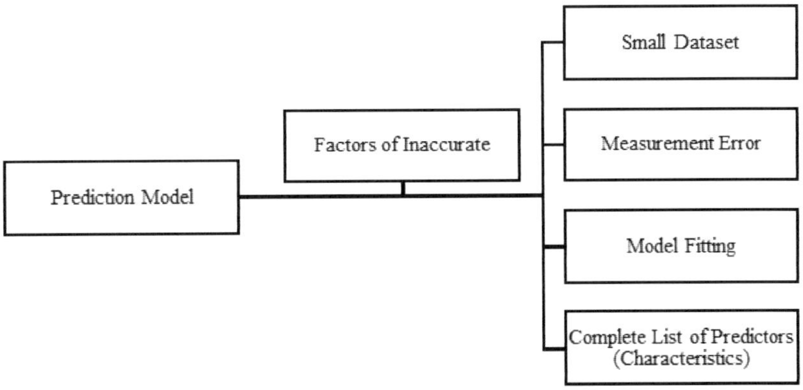

Figure 1. Factors Effecting Inaccuracy of Prediction Model [5]

An effective way is needed to improve the learning accuracy and at the same time, a learning stability is gained. From the literature, one of the effective approach to deal with this problem is adding artificial data to the system. From previous research works, the accuracy of learning algorithm is improved when the amount of training data examples increases. Therefore, this paper is presented to review artificial data generation approach which contain several techniques that robustly handle a small dataset problem for prediction model.

The remainder of this paper is organized as follows. In section 2, related work in small dataset problem is reviewed. In section 3, the artificial data generation approach is explained. Next section will review the general concept of artificial data generation approach to the prediction model. Then this paper is concluded in section 5.

2. Characteristics of Small Data

Data can be defined as a collection of raw facts that has not yet been processed. Then from the processed data, a meaningful form is derived which defined as information. However, the information in small data size is scarce and have some learning limit. The main reason why small dataset cannot provide enough information is that gaps are existed between samples. The gaps between sample and observations are called the information gaps, which cause most of the learning tools are difficult to predict [5]. The Fig.2 shows the example of the information gap that are existed between samples.

According to Vapnik [6], a sample size is considered as small if the ratio of the number of training sample to the Vapnik- Chervonenkis (VC) dimensions of a learning machine is less than 20 [8-10]. VC dimensions is a measure of the capacity for a classificatory, defined as the cardinality of the largest set of points that the algorithm can shatter[8]. However, these theories focus on general machine learning with a large number of training samples, which cannot be applied to practical cases with the small data set learning model [7].

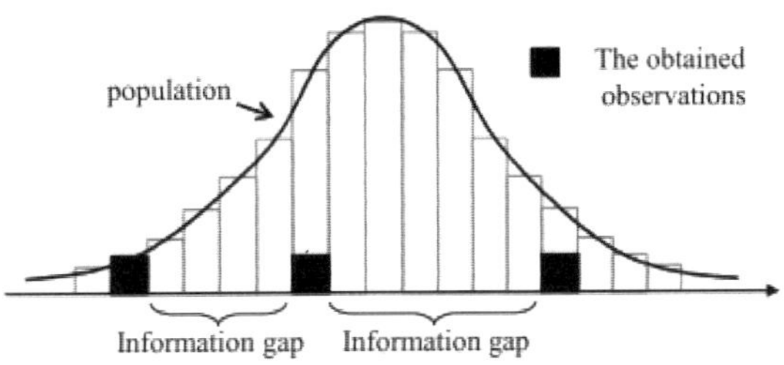

Figure 2. The distribution of a small dataset relative to its population [6]

3. Artificial Data Generation Approach

3.1. Fuzzy Theories

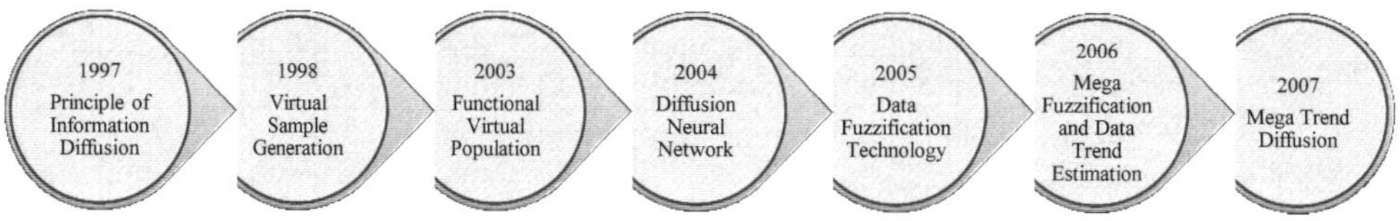

Figure 3. Chronological of Techniques employ Fuzzy Theory

Enhancing learning process in machine learning is important to make sure the performance is improved. One of the approach to enhance the learning process is by adding some synthetic data to the system[9]. Based on the literature, artificial data generation approaches employ some fuzzy theories and neural network techniques which are used to estimate or approximate functions to generate more sample for training sets.

The concept of synthetic data was originally proposed by Niyogi [10] in pattern recognition work which is to improve a 3-D view recognition of an object by increasing the amount of sample data using the prior knowledge gained from small training set. When the amount of the synthetic data is added, it is found that, a 3-D view recognition of an object is improved. However, the estimation of the data effect become the priority in generating a synthetic data. Hence, in the method of a diffusion neural network (DNN) introduced by Huang and Moraga [11], they applied the principle of information diffusion by Huang[12]. Huang proposed a solution for small sample size problem by using fuzzy theories. This is because the incompleteness of data can precisely be represented using fuzzy membership function to represent the similarities between samples. Principle of information diffusion has presented a concept that new artificial examples were used to fill the information gaps which was caused by data incompleteness. The unique diffusion formula is as follows:

$$x_i = u_i \pm \sqrt{-2 \times h \times \ln(\varphi(x_i))} \tag{1}$$

DNN is formed by the principle of information diffusion and traditional neural network. In neural network, the information is processed by assuming two conditions which are 1) the patterns are compatible and 2) the learning pattern for training neural network is sufficient. Otherwise, it is difficult to identify a non-linear system or there will exist a non-negligible error between the real function and estimated function from a trained network [11]. So, the information diffusion is implemented in this technique to derive more pattern to partly fill the gaps between samples which later can produce a better result in identifying non-linear function.

Flexible Manufacturing System (FMS) is a scheduling system which commonly used in production planning and controlling processes. The purposes are to reduce the cost and balance the machine loads. However, it is difficult to decide because the data is not enough for learning process. Li [13] invented the first method named Functional Synthetic Population (FVP) to increase the accuracy of machine learning for FMS scheduling. By following the synthetic data concept, FVP will generate synthetic samples to improve the FMS scheduling knowledge in small data sets learning. The main contribution of this method is the algorithm was developed to expand the domain of the system attributes and added some synthetic samples since for constructing early new scheduling knowledge by adopting neural network algorithm. Since neural network needs enough training set for the learning process, the scheduling knowledge is improved by the added sample. Unfortunately, FVP require many steps to complete [7].

Later, Li employed a data fuzzification technology [13] concept to expand the small dataset obtained to further improve learning accuracy. In this method, the approaches used include data-fuzzifying, domain range expansion, and adaptive-network-based fuzzy inference systems (ANFIS) as neuro-fuzzy learning. After input data were fuzzified, the domain range expansion technique was administered to enhance the domain range allowance in a dynamic and non-enough data environment. The results indicate that the learning accuracy under this strategy is significantly better than that of a traditional crisp data neural networks. However, their proposed method has limitations of the domain range which need to be pre-determined and it must consider the data behavior.

Subsequently, the extended strategy to improve the accuracy of FMS scheduling in small dataset problem is by introducing a new method named Mega Fuzzification and Data Trend Estimation. Because of the limitation existed in previous technique, a new technique is developed by Li [1] intends a combining data fuzzification, data trend estimation, and ANFIS. To build a precise and useful knowledge, considerable quantities of data sets are needed before proceed to the learning process. So, mega-fuzzification or also called as data fuzzification method is presented to determine the possible coverage of a data set as shown in Fig.4.

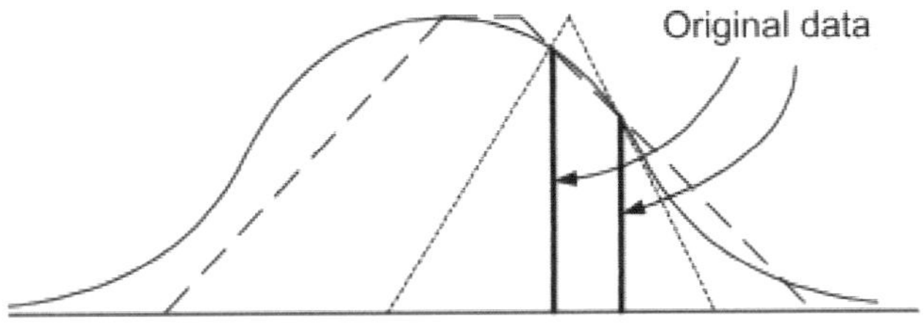

Figure 4. Determining possible coverage of data set using fuzzifying technique [1]

Asymmetric domain range expansion is utilized to estimate the data trend as shown in Fig.5, while expanding the data domain ranges. According to the author, there are some of the matter that should be further studied such as determining membership function type and appropriate height for min and max in the triangular [1]. Besides, the height of min and max in the triangular did has a direct relationship with the attribute domain range estimation. Otherwise, it would lead to either over-estimating or underestimating ranges which cause poor learning process. Thus, to determine the domain ranges of the attribute, providing a suitable method is important.

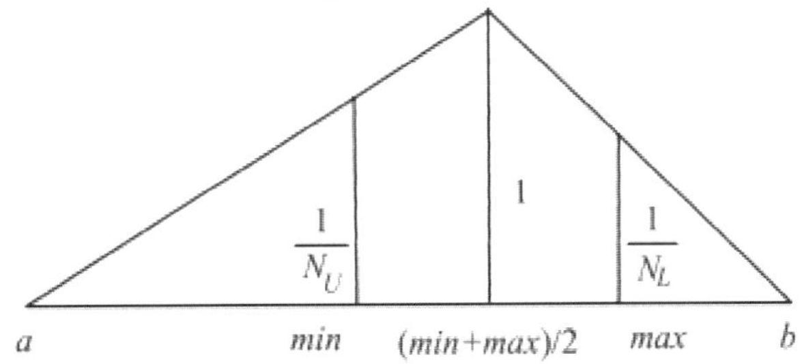

Figure 5. Data Trend Estimation [1]

Later, the Mega Trend Diffusion (MTD) is proposed by Li [3] which employs data trend estimation concept. The mega diffusion produced mega trend diffusion which aimed to avoid over-estimating. The main idea of MTD function is to generate synthetic samples. In statistics view, a normal distribution is necessary before the data analysis is done. However, when the dataset is small, it is often difficult to show the data following the normal distribution [14]. Alternatively, Li [3] used the membership function in fuzzy set theory to calculate the possibility values of synthetic samples than the probability in statistics to avoid the normal distribution assumption. Fig.6 shows the concept of the fuzzy theorem applied to the MTD function.

The conventional method (principle of information diffusion) diffuse each sample individually whereas the MTD method diffuses a set of data using a common diffusion function. MTD method employs information diffusion based on fuzzy possibility distribution to determine possible data set coverage on an attribute domain. Fig. 1 shows the concept of applying the fuzzy theorem to this context. The triangular shape is the membership function, and values a and b are the boundary of the MTD function. The heights of samples m and n are the possibility values of the membership function, and the possibility value is in the interval 0–1. Given a sample set $X = \{x_1, x_{2,}, \dots, x_n\}$, the boundaries a (lower boundary) and b(upper boundary) are defined as below:

$$a = u_{set} - Skew_U \times \sqrt{-2 \times \frac{\hat{s}_x^2}{N_U} \times \ln(10^{-20})} \tag{2}$$

$$b = u_{set} + Skew_L \times \sqrt{-2 \times \frac{\hat{s}_x^2}{N_U} \times \ln(10^{-20})} \tag{3}$$

where u_{set}= (min + max) / 2, with min and max being minimum and maximum values of observation, respectively, $\hat{s}_x^2$is the variance of observations, and ln (10^{-20}) is the diffusion coefficient. N_L and N_U denoted as number of observations which are smaller and greater than u_{set} respectively, and $Skew_L$= N_L / (N_L + N_U) and $Skew_U$= N_L / (N_L + N_U) for the skewness of the possible distribution of the data.

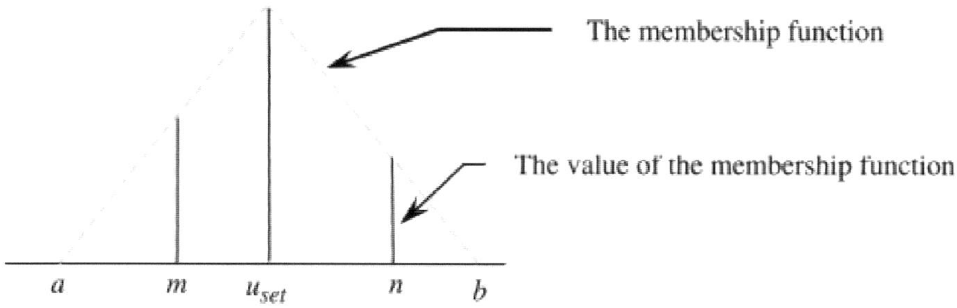

Figure 6. MTD function [3]

Table 1. Computational Result

Technique	Number of Training Set	Learning Accuracy (%)
Mega Fuzzification and Data Trend Estimation [1]	5	69.3
	10	76.7
	20	80.7
	30	88.3
	40	92.2
	50	93.8
	100	94.7
Mega Trend Diffusion [3]	5	78.23
	10	88.53
	20	89.08
	30	91.74
	40	93.41
	50	94.33
	100	95.33

Based on the computational result on Table 1, the learning accuracy of MTD method showed that it is a superior technique and significant to be applied in the real world. The learning accuracy for 30 samples in Mega Fuzzification and Data Trend Estimation are equal to learning accuracy of 10 samples in Mega Trend Diffusion. As a result, the MTD method is widely used in many application including [17-19]. The main idea of MTD is to generate a synthetic sample. However, MTD did not consider the relation among attributes but only for independent attributes. From the finding, one of the improvement is by considering the relationship between attributes. The following section introduce some of the improved technique of MTD function.

3.1.1. Heuristic Algorithm: Genetic Algorithm.
Genetic Algorithm (GA) is one of the optimization algorithm. Even DNN and MTD method discussed previously could solve the small data learning problem by synthetic sample approach, they only considered the data for independent attribute only. Thus, a genetic algorithm was applied which aim to consider the relation of integrated effects and constraints of data attributes [16]. Three main steps were involved and GA was applied in the second step to generate the number of most feasible synthetic samples. The proposed method showed better performance in prediction than without synthetic sample.

3.1.2. Data Transformation.
Data transformation is introduced to bring the data close to the normal distribution to generate a synthetic sample since the small data size is difficult to follow a normal distribution [17]. Although there a many transformation developed, Li introduced Johnson Transformation which is suitable for large data sets (n>30). To make it suitable for small data size, Li introduced Small Johnson Data Transformation (SJDT) function [14] dedicated for small datasets. Johnson system comprises of a family of three curves: bounded system (SB), the log-normal system (SL) and unbounded system (SU) and is represented as follow:

$$Z = \gamma + \eta \times k_i(x, \lambda, \varepsilon) \quad (4)$$

where Z is a random variable that obeys to the standard normal distribution and $k_i(x, \lambda, \varepsilon)$ is the function chosen to cover a wide range of possible occurrence [14]. The function of SB, SL, and SU are defined as follow:

$$\text{Bounded System (SB): } y = \gamma + \eta * \text{Ln}\left(\frac{x-\varepsilon}{\lambda+\varepsilon-x}\right) \quad (5)$$

$$\text{Log-normal system (SB): } y = \gamma + \eta * \text{Ln}\left(\frac{x-\varepsilon}{\lambda}\right) \quad (6)$$

$$\text{Unbounded system (SU): } y = \gamma + \eta * \sin^{-1}\left(\frac{x-\varepsilon}{\lambda}\right) \quad (7)$$

where y is a transformed data, x is a raw data and $(\varepsilon, \gamma, \eta, \lambda)$ are the Johnson parameters. Following to Slifker and Shapiro [18] the parameter of three curves above is estimated based on sample quantiles x_{-3z} , x_{-1z}, x_{1z} and x_{3z} . However, Johnson transformation is suitable for the large dataset (n>30). So, to make Johnson Transformation is fit for a small dataset, the author redefines the four sample quantiles using the Eq.2 and Eq.3 in MTD function. In the proposed SJDT function [14], it is found that the 97% of data is located within interval $[x_{-3z}, x_{3z}]$ when z=1 which is similar to MTD function about 99.73% data. Therefore, the boundaries of a and b in MTD function is used to replace the data interval x_{-3z} and x_{3z}. The four quantiles are now defined as:

$$x_{-3z} = a = u_{set} - Skew_U \times \sqrt{-2 \times \frac{\hat{s}_x^2}{N_U} \times \ln(10^{-20})} \quad (8)$$

$$x_{-1z} = \min \quad (9)$$

$$x_{1z} = \max \quad (10)$$

$$x_{3z} = b = u_{set} + Skew_L \times \sqrt{-2 \times \frac{\hat{s}_x^2}{N_U} \times \ln(10^{-20})} \quad (11)$$

To differentiate the type of transformation action to perform, the selection criteria below are follow:

$$\frac{mn}{p^2} > 1 \text{ for all SU distribution}$$
$$\frac{mn}{p^2} < 1 \text{ for all SB distribution}$$
$$\frac{mn}{p^2} = 1 \text{ for all SL distribution}$$

where $m = x_{3z} - x_{1z}, n = x_{-1z} - x_{-3z}, p = x_{1z} - x_{-1z}$.

Later, the corresponding parameter $(\varepsilon, \gamma, \eta, \lambda)$ are estimated according to three distribution below:

The parameter estimated for SU Distribution are:

$$\eta = \frac{2 \times z}{cosh^{-1}\left[\frac{1}{2}\left(\frac{m}{p}+\frac{n}{p}\right)\right]}, \eta > 0$$

$$\gamma = \eta \times sinh^{-1}\left[\frac{\frac{n}{p}-\frac{m}{p}}{2\left(\frac{m}{p}\times\frac{n}{p}-1\right)^{1/2}}\right]$$

$$\lambda = \frac{2p\left(\frac{m}{p}\times\frac{n}{p}-1\right)^{1/2}}{\left(\frac{m}{p}+\frac{n}{p}-2\right)\times\left(\frac{m}{p}+\frac{n}{p}+2\right)^{1/2}}, \lambda > 0$$

$$\varepsilon = \frac{x_{1z} + x_{-1z}}{2} + \frac{p \times \left(\frac{n}{p}-\frac{m}{p}\right)}{2\left(\frac{m}{p}+\frac{n}{p}-2\right)}$$

The parameter for SB distribution are:

$$\eta = \frac{z}{cosh^{-1}\left\{\frac{1}{2}\left[\left(1+\frac{p}{m}\right)\left(1+\frac{p}{n}\right)\right]^{1/2}\right\}}, \eta > 0$$

$$\gamma = \eta \times sinh^{-1}\left\{\frac{\left(\frac{p}{n}-\frac{p}{m}\right)\left[\left(1+\frac{p}{m}\right)\left(1+\frac{p}{n}\right)-4\right]^{1/2}}{2\left(\frac{p}{m}\frac{p}{n}-1\right)}\right\}$$

$$\lambda = \frac{p \times \left\{\left[\left(1+\frac{p}{m}\right)\left(1+\frac{p}{n}\right)-2\right]^2-4\right\}^{\frac{1}{2}}}{\frac{p}{m}\frac{p}{n}-1}, \lambda > 0$$

$$\varepsilon = \frac{x_{1z} + x_{-1z}}{2} - \frac{p}{2}\left(\frac{\frac{m}{p}+1}{\frac{m}{p}-1}\right)$$

and the parameters estimated for SL distribution are:

$$\eta = \frac{2z}{\ln\left(\frac{m}{p}\right)}$$

$$\gamma = \eta \times \ln\left[\frac{\frac{m}{p} - 1}{p\left(\frac{m}{p}\right)^{1/2}}\right]$$

$$\varepsilon = \frac{x_{1z} + x_{-1z}}{2} - \frac{p}{2}\left(\frac{\frac{m}{p} + 1}{\frac{m}{p} - 1}\right)$$

3.2. Resampling Mechanism

Instead of using fuzzy theories to increase the amount of observation, the statistical method is also employed as an alternative approach. Bootstrap is the most well-known synthetic sample generation method developed based on statistical resampling mechanism. Bootstrap randomly resample the original observations with replacement [7] which is differ from previous techniques which applies fuzzy theories. Bootstrap is introduced by Efron [21] in early 1979. However, it was not quite favorable because, according to Tsai and Li [5], bootstrap aimed to increase the amount of obtained observation rather than filling the information gaps of new observations. Although it is unfavorable technique, the learning accuracy increase in the prediction of bladder cancer cell for example [7].

3.3. Data Clustering

From the point of fuzzy membership generation, the fuzzy c-means (FCM) is similarly to MTD method. The difference is just in the calculation procedure. FCM is a data clustering technique proposed by Bezdek which applied fuzzy approach in clustering procedure. Generally, the FCM algorithm allows one piece of data to belong to different cluster with different membership degree of each cluster. Sezer used FCM as a method to produce synthetic training set for small dataset problem [22]. The author [22]found that most of the previous studies used the methods which are hard to understand. So, a better and easy method is more preferred to represent the information within the data. Hence, FCM is chosen as the alternative method to produce a synthetic data using centroid calculation scheme of FCM to reflect all the closeness relationships within the data. The instance is considered to have some closeness with another instance in another cluster if it is in the boundary of the cluster. Thus, centroid calculation scheme will identify whether the data is a good candidate to be synthetic data based on the centroid of the classes in different random samplings [22]. FCM need a specification of clusters number at the beginning. Each of cluster centroid is then randomly initialized. The iterative process is followed for computing the centroid of each cluster by using centroid calculation in Eq.12 until convergence criteria are met.

$$ce_i = \frac{\sum_{j=1}^{N}\left(u_{c_i}(x_j)\right)^m . x_j}{\sum_{j=1}^{N}\left(\mu_{c_i}(x_j)\right)^m} \quad (12)$$

where $u_{c_i}(x_j)$ is the membership degree of the *j*th instance "*x*" to cluster "*c*", *N* is the number of instances and ce_i is the centroid of *i*th cluster. After that, the membership of each instance to the cluster is calculated by using Eq.13.

$$u_{c_i}(x_j) = \frac{1}{\sum_{k=1}^{C}\left(\frac{\|x_j - ce_i\|}{\|x_j - ce_k\|}\right)^{2/(m-1)}} \quad (13)$$

where "c" is the total of cluster number, ce_k is the centroid of the *k*th cluster while "*m*" is the degree of fuzziness.

4. Artificial Data Generation Concept to Prediction Model

From the findings, increasing the number of synthetic samples can increase the learning accuracy of a model. This is because the gaps between samples are filled with artificial data generation approach as discussed previously. We can conclude the overall concept of artificial data generation approach in small dataset is:

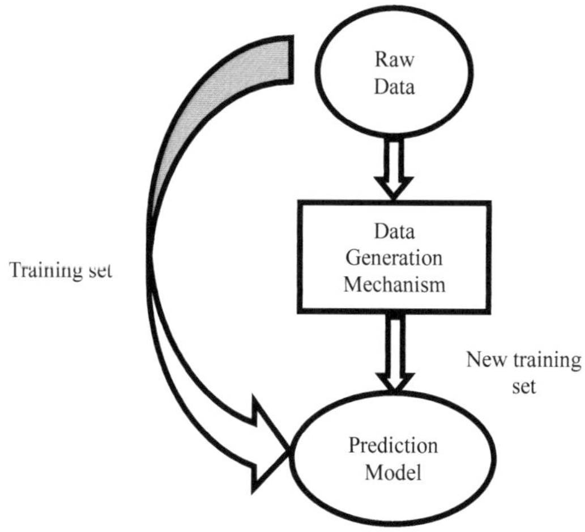

Figure 7. Data Generation Concept in Prediction Model

Referring to Fig.7, the concept generally can be explained from the data obtained, the data generation mechanism increases the number of training set to be used in the learning process to produce a model. The new form of training set then is combined with the original data to be used as a training sample of a model.

5. Conclusion

This paper presented a review on small dataset problem occur in certain areas where data is hard to obtain. Small dataset problem which cause poor learning performance and significant uncertainty to prediction model, is the reason why an efficient way is needed to build a precise prediction model. From literature, artificial data generation approach is one of the effective methods to cope with small dataset problem. Different techniques have been introduced to overcome a small dataset problem. Based on the techniques discuss earlier, MTD technique is widely used as a synthetic sample generation method to increase the amount of training sample. Besides, the improvement of MTD method has also been explored and introduced.

Acknowledgement

This work is funded by the Ministry of Higher Education Malaysia and Universiti Teknikal Malaysia Melaka (UTeM) through the PJP High Impact Research Grant [PJP/2016/FTMK/HI3/SO1473]

References

[1] D. C. Li, C. Sen Wu, T. I. Tsai, and F. M. Chang, "Using mega-fuzzification and data trend estimation in small data set learning for early FMS scheduling knowledge," *Comput. Oper. Res.*, vol. 33, no. 6, pp. 1857–1869, 2006.

[2] D. C. Li, C. C. Chang, C. W. Liu, and W. C. Chen, "A new approach for manufacturing forecast problems with insufficient data: The case of TFT-LCDs," *J. Intell. Manuf.*, vol. 24, no. 2, pp. 225–233, Jul. 2013.

[3] D. C. Li, C. Sen Wu, T. I. Tsai, and Y. S. Lina, "Using mega-trend-diffusion and artificial samples in small data set learning for early flexible manufacturing system scheduling knowledge," *Comput. Oper. Res.*, vol. 34, no. 4, pp. 966–982, 2007.

[4] A. M. Al-bakary and S. H. Ali, "Genetic Programming Data Construction Method to Handle Data Scarcity Problem," no. February, pp. 1–10, 2010.

[5] C.-H. Tsai and D.-C. Li, "Improving Knowledge Acquisition Capability of M5' Model Tree on Small Datasets," *2015 3rd Int. Conf. Appl. Comput. Inf. Technol. Int. Conf. Comput. Sci. Intell.*, pp. 379–386, 2015.

[6] V. N. Vapnik, *The Nature of Statistical Learning Theory*, vol. 8, no. 6. 2000.

[7] G. Chao, T. Tsai, T.-J. Lu, H. Hsu, B. Bao, W. Wu, M. Lin, and T. Lu, "A new approach to prediction of radiotherapy of bladder cancer cells in small dataset analysis," *Expert Syst. Appl.*, vol. 38, no. 7, pp. 7963–7969, 2011.

[8] R. Andonie, "Extreme data mining: Inference from small datasets," *Int. J. Comput. Commun. Control*, vol. 5, no. 3, pp. 280–291, 2010.

[9] T. I. Tsai and D. C. Li, "Utilize bootstrap in small data set learning for pilot run modeling of manufacturing systems," *Expert Syst. Appl.*, vol. 35, no. 3, pp. 1293–1300, 2008.

[10] P. Niyogi, "niyogi-poggio-ProcIEEE-1998.pdf," *Proc. IEEE 86(11)*, 1988.

[11] C. Huang and C. Moraga, "A diffusion-neural-network for learning from small samples," *Int. J. Approx. Reason.*, vol. 35, no. 2, pp. 137–161, 2004.

[12] C. Huang, "Principle of information diffusion," *Fuzzy Sets Syst.*, vol. 91, no. 1, pp. 69–90, 1997.

[13] D. C. Li, C. Wu, and F. M. Chang, "Using data-fuzzification technology in small data set learning to improve FMS scheduling accuracy," *Int. J. Adv. Manuf. Technol.*, vol. 27, no. 3–4, pp. 321–328, 2005.

[14] D.-C. Li, I.-H. Wen, and W.-C. Chen, "A novel data transformation model for small data-set learning," *Int. J. Prod. Res.*, vol. 7543, no. June, pp. 1–11, 2016.

[15] D.-C. Li, W.-T. Huang, C.-C. Chen, and C.-J. Chang, "Employing virtual samples to build early high-dimensional manufacturing models," *Int. J. Prod. Res.*, vol. 51, no. 11, pp. 3206–3224, 2013.

[16] D. C. Li and I. H. Wen, "A genetic algorithm-based virtual sample generation technique to improve small data set learning," *Neurocomputing*, vol. 143, pp. 222–230, 2014.

[17] D.-C. Li, C.-C. Chang, and C.-W. Liu, "Using structure-based data transformation method to improve prediction accuracies for small data sets," *Decis. Support Syst.*, vol. 52, no. 3, pp. 748–756, 2012.

[18] J. F. Slifker and S. S. Shapiro, "The johnson system: selection and parameter estimation," *Technometrics*, vol. 22, no. 2, pp. 239–246, 1980.

[19] B. Efron, "Bootstrap Methods: Another Look at the Jackknife," *The Annals of Statistics*, vol. 7, no. 1. pp. 1–26, 1979.

[20] E. a. Sezer, H. a. Nefeslioglu, and C. Gokceoglu, "An assessment on producing synthetic samples by fuzzy C-means for limited number of data in prediction models," *Appl. Soft Comput. J.*, vol. 24, pp. 126–134, 2014.

Internet of Things in Higher Education: A Study on Future Learning

Hanan Aldowah[1, 3], Shafiq Ul Rehman[2], Samar Ghazal[1, 3], Irfan Naufal Umar [1]

[1]Universiti Sains Malaysia (USM), Centre for Instructional Technology and Multimedia (CITM), Penang, Malaysia
[2] National Advanced IPv6 Centre (IPV6), Universiti Sains Malaysia (USM), Penang, Malaysia
[3]Centre for Information System and Technology (CIST), Hodiedah University, Hodiedah, Yemen

hanan_aldwoah@yahoo.com

Abstract. In the coming years, technology will impact the learning experience in many ways. Internet of Things (IoT) continues to confirm its important position in the context of Information and Communication Technologies and the development of society. With the support of IoT, institutions can enhance learning outcomes by providing more affluent learning experiences, improved operational efficiency, and by gaining real-time, actionable insight into student performance. The purpose of this study is to find out the potential of IoT in higher education and how to maximize its benefits and reducing the risks involved with it. Further efforts are necessary for releasing the full potential of IoT systems and technologies. Therefore, this paper presents a study about the impact of IoT on higher education especially universities. IoT stands to change dramatically the way universities work, and enhance student learning in many disciplines and at any level. It has huge potential for universities or any other educational institutions; if well prepared to ensure widespread and successful implementation by leadership, staff, and students. IoT needs development where universities can lead. Academics, researchers, and students are in a unique place to lead the discovery and development of IoT systems, devices, applications, and services. Moreover, this paper provides an evidences about the future of IoT in the higher education during the next few years, which have offered by a number of research organizations and enterprises. On the other hand, IoT also brings tremendous challenges to higher education. Hence, this paper also presents the perspective on the challenges of IoT in higher education.

1. Introduction

Internet of Things (IoT) is the transformation process in numerous aspects of our daily life. IoT technologies differ from previous innovations as they are ubiquitous, and encourage solutions to be intelligent and autonomous [1]. Advances in the IoT are a major strategic technology trend [2]. Ubiquitous sensors and the ability to bridge the gap between the physical world and the machine world were perceived as the conceptual framework for the new learning model. The thinking behind this great paradigm shift is the ability to embed sensors into any object and use Machine-to-Machine (M2M) communication to connect billions of objects/devices to the current Internet infrastructure. The entirety of the physical world is coming online rapidly.

IOP Conf. Series: Journal of Physics: Conf. Series **892** (2017) 012017 doi:10.1088/1742-6596/892/1/012017

IoT is developing quickly and becoming an increasingly growing topic that creates excitement and anxiety around the world [3]. There are plenty of indications showing that the IoT will to change many sectors, including higher education institutions, especially universities. Now, universities has an opportunity to lead the technical development and the innovations models for the IoT, and to build the leaders of the IoT into the future, as well as to address the TIPPSS risks which stands for Trust, Identity, Privacy, Protection, Safety, and Security related to the IoT.

The IoT is a global physical network which connects devices, objects and things to the Internet infrastructure to communicate or interact with the internal and the external environment as illustrated in Figure 1, and for the purpose of exchanging information through the information sensing devices according to specific protocols. Thus, IoT is enabling connectivity for anything and for anyone to be networked around the world anytime, and anywhere using any network or any service [4] to achieve the goal of intelligent identifying, tracking, and managing things [5]. It is an extension and expansion of Internet-based network, which expands the communication between human to human (H2H), human to things (H2T)or things to things (T2T) [6] as presented in Figure 2.

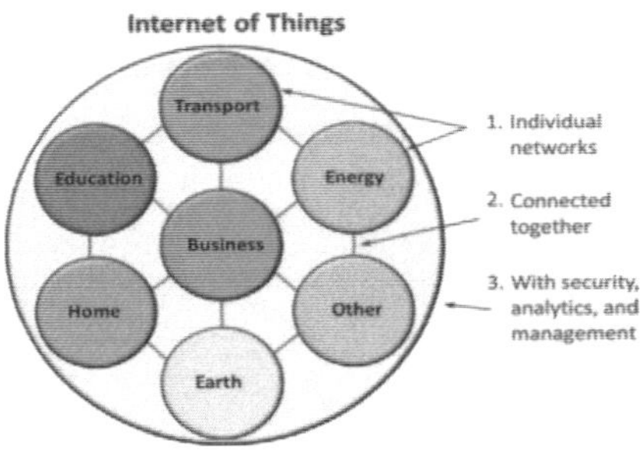

Figure.1. IoT viewed as a global network.
(Source: Cisco IBSG, April 2011).

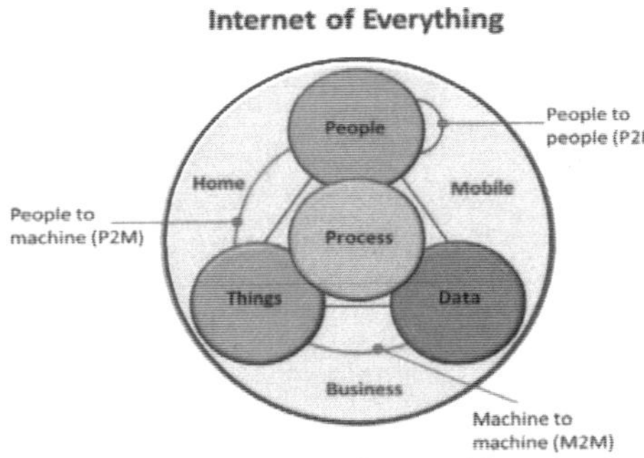

Figure.2. Internet of everything
(Source: Cisco, April 2011).

A number of research organizations and analysts have predicted about the future of IoT and its potential impact on the Internet: billions of physical devices, all over the world, that have digital sensors and are interrelated by leveraging any network. According to the research conducted by Juniper [7], an estimated 13.4 billion devices were connected in 2015, which represent more than the total of population on the earth at the time, and this number is expected to triple to 38.5 billion devices by 2020. The opportunity of interacting with lots of everyday objects connected to the Internet allows individual to access unlimited information anytime and anywhere. This vision opens a new horizon of ideas and developments that is already being considered by research scholars and academics.

The IoT vision is grounded in the belief that the firmly fixed advances in microelectronics, communications and information technology we have witnessed in current years will endure into the predictable future [8]. Applications for the IoT are already being leveraged in sectors like healthcare and customer service. Now, universities and schools are joining the party. Some of the ways the IoT can benefit education may be manifest, while others are not as obvious. So this paper will present the biggest implications for connected devices in higher education and how they could shape the learning for the next generation.

The future of universities is not about using and employing the available technology. It is about how universities will adapt to the changing needs of the future knowledge worker, the future of work, and the economy. This paper presents an overview of IoT in higher education institutions, especially in universities and takes a look at several emerging trends that are evolving higher education, and explore the potential impact of IoT and the future of the IoT in higher education. In addition, exploring some of IoT challenges regarding higher education sector.

2. Basic Components of IOT

The IoT is empowering technology researchers to develop smaller and more affordable wireless systems that consume less power and can be integrated into almost any type of device [9]. There are three IoT components which enable seamless connections which are: Hardware: made up of sensors, actuators and embedded communication hardware, Middleware: on demand storage and computing tools for data analytics, and Presentation: novel easy to understand visualization and interpretation tools which can be widely accessed on different platforms and which can be designed for different applications [10]. There are a number of potential approaches for introducing low-power communications to an IoT node, ranging from purpose-designed protocols such as ZigBee to low-power variants of Bluetooth, Wi-Fi and NFC. Although Wi-Fi is the most popular form of integrated wireless technology and the best power-per-bit transmission efficiency, IoT enhances other formats including Radio Frequency Identification (RFID) technology which is used throughout business, industry and personal technology systems and enables design of microchips for wireless data communication [10]. Some of this technology can add wireless sensor capabilities (WSN) to any type of device, like FitBit wearable fitness trackers, and books.

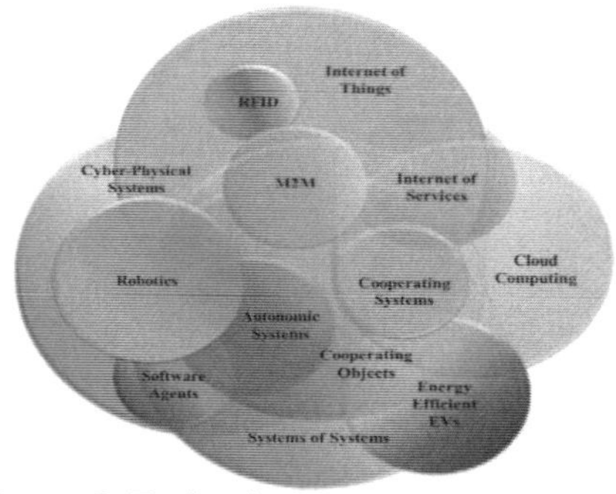

Figure 3: Technology convergence.
Source: (Friess, 2013)

3. Concept of Digital Campus

Digital Campus System is an important platform for students to get all kinds of information [11]. New technologies are also affecting other areas of campus administration. There is an increasing demand for higher education institutions, especially, universities to digitize their content and activities, and adapt their methods to allow academic and researchers to work effectually in a digital environment [12]. A well designed physical campus, completely integrating technology, is fundamental for building the brand of digital university by enhancing the student experience, and providing the appropriate settings and facilities for teaching, learning and research. It promotes, supports and encourages lifelong learning [13]. A digital university must have the technology that enabled teaching and learning, and empower collaborative research. All contemporary digital threats can be faced by universities if they compete, however few have the vision, flexibility, platforms, or appropriate leadership, to put in place the strategies to ensure that they can innovate, or react to marketplace conditions.

Within a digital campus, technology can reduce operational costs, improve security, and offer tools for researchers, academics, students and staff. These benefits provide real value to university operations and developments, the experience of students, and researchers. The digital campus comprises two main components. Firstly, it reuses the IT Service Delivery Platform – end-to-end infrastructure to provide network connectivity, mobility and security for all applications and services across the campus. Secondly, it includes a large number of Internet of Things (IoT) applications operating over the platform system to support the professional of the university, enable teaching and learning activities, and enhance student's experience. According to Cisco -"Digitizing Higher Education To enhance experiences and improve outcomes"- IoT applications differ from conventional network applications as they support sensors and sensor data, rather than users and user data. IoT applications for the digital campus include five main categories: Building Control and Management; Security and Access Control; Video and Information Systems; Location and Attendance Systems; Energy Monitoring and Control.as shown in Figure 4:

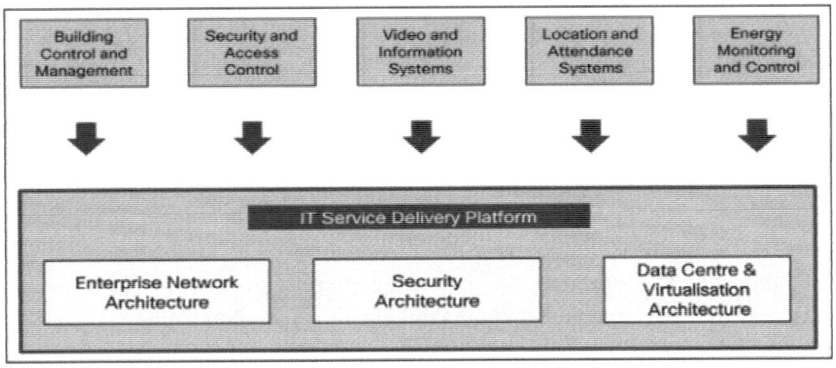

Figure 4: IoT Applications for the Digital Campus
Source: Cisco- Digitizing Higher Education To enhance experiences and improve outcomes.

The wireless network has a main role to play within the digital campus, thus must be designed to meet the high demands of a modern university.

In addition, IoT reform and change the teaching and learning in the campus, for example, the IoT in the training of the same sight, making the teaching space, training venues, sports venues, learning dormitories, restaurants and students across the campus of IoT, making the campus to train students anytime and anywhere, can become a physical, mental, and skills place to acquire a full range of learning and training. This will lead colleges to become ubiquitous learning and training [14].

4. Impact of Internet of Things on Higher Education

The IoT is going to affect every part of society at some point in the near future. Higher education institutions in general, and universities in particular, can work across disciplines and lead the progress of the IoT technologies, business models, ethics, and leaders of the IoT enabled economy of the future. For instance, university instructor of computer science and engineering are directing IoT labs for the development of IoT technologies. In addition, Informatics College can teach how to leverage the sizes of IoT data, with TIPPSS. Also, they can work with business colleges to set and design IoT courses to create new business models. Medical colleges can empower the Internet of Medical Things as well as, Law colleges can teach IoT ethics, privacy, and policy. According to Zebra technologies, as higher education institutions commence to develop and leverage solutions such as radio frequency identification (RFID) and cloud computing through IoT technologies, they will be able to analyse and manage Big Data.

The IoT is not just a technology update and development within the industry, but can lead to expand the change to the whole society including higher education institutions. IoT will lead the change and reform the higher education institutions. According to [14], IoT will lead to changes in educational technology, reform the education, change in teaching, change in learning, management of change, experimental and practical changes, changes in campus, teaching resources changes and others.

With the development of IoT, the prospective application in higher education lies in the three aspects: students' progressive evaluation, integration of current teaching platforms and development of educational middleware [15]. This change provides increased convenience for students, and makes the teaching process more effective for instructors and professors. The flow in connected devices and technology means that instructors and professors can focus on the actual learning that is more useful to the students rather than perform the routine task.

In addition, IoT has the ability to increase the learning experience by providing for real-time and actionable insights into student performance. Nowadays, students particularly in university are gradually moving away from textbooks to new technologies such as tablets and laptops. The advanced e-learning applications allow students to learn at their own pace and have an identical learning experience in classrooms and homes [16], which rises progression and satisfaction rates as well as instructors can deliver one-to-one instruction and persistent student assessments [17]. Moreover, through IoT technology, professors can collect data about students' performance and then determine which ones need more care and attention. This data analysis also helps instructors accurately change plans and methods for future classes. Additionally, connected devices can allow instructors to do dynamic classroom. Interventions as well as logging attendance will be simplified if students have a wearable device that tracks ECG patterns. Furthermore, these devices can redirect a student's attention by giving a warm up activity and exercise to do on their own devices. Also, EEG sensors can be used during courses to monitor students' cognitive activities.

This vision and understanding give stakeholders with a view of students, organization, and financial assets. This asset intelligence enables organization to make informed decisions in order to enhance student knowledge and learning experiences, operational proficiency, and the security of campus. According to Zebra technologies (2015), by enhancing asset intelligence, educational institutions can enhance outcomes by adding values in some areas include: Enhanced Learning Experiences and Outcomes, Improved Operational Efficiency, Safer Campus Designs.

Moreover, outside of the classroom, universities can use connected devices to monitor their students, staff, resources and equipment at a reduced operating cost [18]. Furthermore, the growth of mobile technology and the IoT enable universities to improve the security of campuses, enhance access to information and applications at anytime from anywhere, and keep track of main resources [10]. IoT is changing the student learning experience besides facilities management by connecting individual, data and things.

5. Future of IoT in Higher Education

Universities have long realized the ability of technology to disrupt teaching, learning, and assessment. Furthermore, technology disruption is fundamental if a modern university is to distinguish its student offer, so increasing admissions, improving retention, and delivering desired outcomes. But preparing students to be confident for the world of work is complex. It requires strong academic leadership, access to a high quality curriculum and content, and the exposure of students to the effective use of new technology. With the development of IoT, many institution of higher education have started to focus on the related technology and application of the IoT [9, 19]. This attempt is also used in university [20]. The Internet has deeply rooted itself into colleges and universities, and e-learning has become common practice in most universities systems [16]. Although it is not an obvious application of the IoT, however, education is on that list [10] and the applications of the IoT in universities are numerous, and the implications for this are massive. IoT will allow for better operational efficiency in all learning environments. IoT can support classroom instruction by improving learning setting, enhance learning resources, improve methods and techniques of learning, raise management efficiency, and save management costs. The resources available for learning on devices, like e-books, are more engaging and interactive. However, there is a constant need for new technologies for learning process, for instance, high-speed wireless networks with the bandwidth for streaming audio and video lessons.

According to the Citrix 2020 Technology Landscape Report (2015), in the next five years, IoT technology will enhance the learning experience in different ways. Learning experience will continue to become more virtual, students will consume knowledge and learning in new ways, and classrooms will be better equipped for learning. Eventually, learning will become an amazing experience for instructors and students with knowledge accelerating while bringing new ideas and solutions around the world. As well, students are prepared for the future of work and expectations at the workplace of the future.

Technology will always have a place in all educational disciplines [21]. IoT also has many opportunities for Science, Technology, Engineering, and Mathematics (STEM) disciplines, such as computer programming and physical computing. It is easy to foresee how IoT capabilities can be used in STEM disciplines, robotics, and anything having to do with collecting specific data. It is all in the potential of the IoT. However, ultimately the educationists need to be able to identify the right technology and integrate it properly in the classroom for learning to evolve. Although main IoT technologies are so far unclear, the point certainly is that a lot of contents are the outcome of new development phase. Considering the demand of more professional research, setting the IoT major is relative easy and applicable for graduate students. However, for undergraduate students, they still need a wide range of basic courses, so it is not easy to set IoT major independently like other majors at present. New training methods shall be explored for undergraduate students [22]. Different colleges have to explore the appropriate approach according to their own characteristics. The system approach and courses content need to be progressively established and improved. Since IoT is achieving the unity of the virtual world and the physical world, many new training methods and cross-cutting areas will be generated in the future [3].

Moreover, the future IoT economy can be shaped by experts and leaders in higher education sector and by educating the students [23]. The development within higher education systems will visualize, improve, and lead the new technology innovations. Therefore, higher education sector must work with business and industrial sectors to shape and build the future of an IoT-enabled economy. Furthermore, higher education sector, especially universities, have the opportunity to lead the future of IoT technologies by designing courses for technical and business leaders and by facilitating students and researchers work to build new business methods that leverage IoT technologies in a multidisciplinary way.

In 2016, a workshop conducted by IEEE, National Science Foundation (NSF), and Internet2, entitled "End to End Trust and Security for the Internet of Things", and followed by an IEEE Experts in Technology and Policy Forum, the experts and participants asserted that the IoT needs innovation and development which researchers, professors, academics ,and students in universities and higher education sector are in a unique place to lead the innovation and development of IoT devices, systems, applications, and services. Also, they emphasized that new platforms and ideas across disciplines must be developed and discovered to solve many problems and issues that we are facing nowadays. In addition, IoT and data analytics tools can be leveraged to develop and improve effectiveness and competences on campus and across societies, to improve information capture, to address security and privacy issues, to minimize the energy use, and to analyze data and provide actionable insights and understanding to develop and enhance health sector. To build this system sight, working across skill sets and disciplines are required. Furthermore, academics and research scholars can build end-to-end TIPPSS solutions for the IoT as well as they can build IoT devices, and services with a "defense in depth" strategy, adding in security at the hardware, software, firmware, and service levels.

6. Challenges of IoT in Higher Education

IoT brings tremendous challenges and opportunities to higher education. The unique growth of ubiquitous computing, developing IoT technologies such as cloud computing, and big data and analytics are helpful not only in improving the core values of teaching and quality of research but also developing an IoT society and encouraging a new digital culture. With increasing online degree opportunities and seamless access to instructional content in both structured and unstructured formats, the IoT leads digital momentum into higher education institutions. IoT is a dramatic shift in the traditional instructional paradigm while integrating broader disciplines, including social science, to enrich the value of big data available from social media. Some of the IoT challenges in higher education sector include:

6.1. Cloud Computing

Many universities are using hybrid cloud as their enterprise architecture for hosting IoT applications. The combination of millennials, the most tech-savvy students in the universities, as well as the rise of tablet and mobile technology, has opened new methods to increase the effectiveness of enterprise architecture, instructional technologies, research and learning environments. With ubiquitous computing, the cloud provides seamless connections and services to information technology services. Presently, enterprise architecture in many higher education institutions depend on hybrid cloud infrastructures with computing platforms on private clouds, while enterprise and instructional applications gradually move to public clouds. Enterprise architecture in these institutions need reduce latency time because of the demand for content in instructional technologies, the huge increase in audio and videos for instructions, and the need for active enterprise networks.

6.2. Instructional Technologies

The growing use of learning management systems LMS like Moodle and Blackboard is creating massive amount of structured and unstructured data such as audio and video content. Sophisticated electronic schoolrooms equipped with lecture capture systems and web streaming provide an opportunity for students to access instructional contents on demand at any time [9] .

6.3. Mobility Applications

IoT applications are being increasingly used to integrate mobile learning applications and for assessment and grading systems. The ideal application can assist students to benefit from learning resources, manage assignments, and work on tasks. Instructors also use some of these applications to teach highly specialized concepts, complex physical, scientific simulations, and social topics.

6.4. Security and Privacy

The implementations of IoT technologies present new and unique security and privacy challenges and issues. Addressing these challenges and issues to ensure security in IoT devices and services should be a fundamental priority [8]. One of the fundamental criteria for IoT is the need to include effectual and trustworthy privacy and security mechanisms [24]. Higher education is vulnerable to the security and privacy of the IoT ecosystem.

Even though there has been further momentum to deal with the security of the IoT infrastructure, there is still no strategy to identify business risks associated with data breaches. Higher education sector need to develop standards to secure IoT applications. As higher education creates millions of future workers, it has to embrace IoT platforms and systems even with the challenges of IoT financing, evolving digital educational pedagogy, training, and interdisciplinary research. In addition, IoT applications must engage the future workforce morally and ethically to address cyber security issues as society depends more on IoT applications. Therefore, a collaborative method to safety and security will be required to develop solutions in effective and appropriate way to face IoT security challenges. Furthermore, the full potential of the IoT depends on strategies that consider people's privacy. Therefore, to fulfill these opportunities, there is need to develop new strategies that consider an individual's privacy choices and expectations, whilst still promote innovation in new technologies and services [8].

6.5. *Research Computing*

Higher education continues to benefit from IoT integration. As the cost of hardware reduces, interdisciplinary research has gained momentum in the last years. In addition, with the availability of big data, even smaller universities can increase their interdisciplinary research footprint and put in high performance computing (HPC), big data platforms, and analytics. STEM education has seen the necessity of identifying broader collaboration with IoT ecosystems by using sensor technologies, Unmanned Aerial Vehicles (UAVs) and microcontrollers. Engineering laboratories use audio video technologies, UAV, Raspberry Pi and open source systems (OSS) that are driving innovations and enhancing learning processes in engineering projects. Social science researchers intrigued by the plethora of big data generated by social media and omnipresent computing are constantly using distributed computing platforms such as HPC, GPU clusters, Hadoop clusters and big data analytics to improve IoT research.

6.6. *Quality and Ethics*

The quality of learning both online and on campus and the rising cost of higher education has been intensely debated in latest years. The IoT offers unique opportunities to deliver digital courses. However, it also introduces challenges to maintain the quality of instruction and evaluation of students' work. IoT educational applications need tools and technologies for instructors, professors and the scientific community to improve the quality of research and address ethics issues within higher education.

6.7. *Financing*

The cost of information technologies continues to increase every year as content and an application. These application stacks continue to grow both horizontally and vertically on instructional technologies, research computing and enterprise technologies. Alongside the information technology and laboratory fees, most universities do not have a strategy for sharing costs and identifying the total cost of ownership for an IoT infrastructure. Higher education must come up with new ideas to finance an information technology infrastructure and services.

7. Conclusion and Future Work

With the advancement in technology i.e. Internet of Things, universities can resolve many challenges such as; keeping track of essential resources, develop access to information, build smarter plans, and design safer campuses. IoT systems have tremendous potential to bring significant values to higher education by engaging and motivating the students and staff, and to increase speed of learning. The purpose of this study was to find out the potential of IoT in higher education and how to maximize its benefits while addressing its challenges and reducing the risks involved with it. Therefore, our future work will be to focus on IoT implementation in higher education.

References

[1] Kahlert, M., *Understanding customer acceptance of Internet of Things services in retailing: an empirical study about the moderating effect of degree of technological autonomy and shopping motivations*. 2016, University of Twente.
[2] High, P.G., *High, P. (2015). Gartner: Top 10 Strategic Technology Trends For 2016. Retrieved from http://www.forbes.com/sites/peterhigh/2015/10/06/gartner-top-10-strategic-technology-trends-for-2016*. 2015.
[3] Ning, H. and S. Hu, *Technology classification, industry, and education for Future Internet of Things*. International Journal of Communication Systems, 2012. **25**(9): p. 1230-1241.
[4] Friess, P., *Internet of things: converging technologies for smart environments and integrated ecosystems*. 2013: River Publishers.
[5] Stankovic, J.A., *Research directions for the internet of things*. IEEE Internet of Things Journal, 2014. **1**(1): p. 3-9.
[6] Chen, S., et al., *A vision of IoT: Applications, challenges, and opportunities with china perspective*. IEEE Internet of Things journal, 2014. **1**(4): p. 349-359.
[7] JuniperResearch, *Internet of Things' Connected Devices to Almost Triple to over 38 Billion Units by 2020*. 2015.
[8] Agarwal, S. and S. Pati, *Study of Internet of Things*. International Journal for Scientific Research & Development, 2016. **4**(05): p. 4.
[9] Jin, D., *Application of" Internet of Things" in Electronic Commerce*. International Journal of Digital Content Technology & its Applications, 2012. **6**(8).
[10] Gubbi, J., et al., *Internet of Things (IoT): A vision, architectural elements, and future directions*. Future Generation Computer Systems, 2013. **29**(7): p. 1645-1660.
[11] Han, Z. and J. Wang, *Development and Research of Digital Campus System Based on Android*. 2014.
[12] Sherson, G., *Education and the Digital Campus* 1999: p. 9.
[13] Porter, A. and Mark Sherwin, *The Digital Campus The Online Future For Higher Education*. 2013: p. 38.
[14] Tianbo, Z. *The internet of things promoting higher education revolution*. in *Multimedia Information Networking and Security (MINES), 2012 Fourth International Conference on*. 2012. IEEE.
[15] Zhiqiang, H. and Z. Junming, *The Application of Internet of Things in Education and Its Trend of Development [J]*. Modern Distance Education Research, 2011. **2**: p. 019.
[16] Aldowah, H., S. Ghazal, and B. Muniandy, *Issues and Challenges of Using E-Learning in a Yemeni Public University*. Indian Journal of Science and Technology, 2015. **8**(32).
[17] Ghazal, S., Z. Samsudin, and H. Aldowah, *Students' Perception of Synchronous Courses using Skype-based Video Conferencing*. Indian Journal of Science and Technology, 2015. **8**(30).
[18] Yan-lin, L.L.-y.Z., *The Application of the Internet of Things in Education [J]*. Modern Educational Technology, 2010. **2**(005).
[19] Fan, S., Z. yu, and H. Guo, *Affects of internet of things on Supply Chain management, China Economics and Trade*. 2009.
[20] Qi, Ai-qin Shen, and Yong-jun. *The Application of Internet of Things in Teaching Management System*. in *Information Technology, Computer Engineering and Management Sciences (ICM), 2011 International Conference on*. 2011. IEEE.
[21] Sundmaeker, H., et al., *Vision and challenges for realising the Internet of Things*. Cluster of European Research Projects on the Internet of Things, European Commision, 2010.
[22] Tan, D., *Engineering Technology, Engineering Education and Engineering Management: Proceedings of the 2014 International Conference on Engineering Technology, Engineering Education and Engineering Management (ETEEEM 2014), Hong Kong, 15-16 November 2014*. 2015: Crc Press.
[23] Kortuem, G., et al., *Educating the Internet-of-Things generation*. Computer, 2013. **46**(2): p. 53-61.

IOP Conf. Series: Journal of Physics: Conf. Series **892** (2017) 012017 doi:10.1088/1742-6596/892/1/012017

[24] Mineraud, J., et al., *A gap analysis of Internet-of-Things platforms.* Computer Communications, 2016. 89, 5-16.

Hybrid Clustering-GWO-NARX neural network technique in predicting stock price

Debashish Das[a,b], Ali Safa Sadiq[a,c], Seyedali Mirjalili[d], A. Noraziah[a]

[a]Faculty of Computer Systems & Software Engineering, University Malaysia Pahang
Kuantan, Pahang Darul Makmur, Malaysia
debashish.das@apu.edu.my, alisafasadiq@ump.edu.my, noraziah@ump.edu.my

[b]Faculty of Computing, Engineering and Technology, Asia Pacific University of Technology and Innovation
TPM, Bukit Jalil, Kuala Lumpur 57000, Malaysia
debashish.das@apu.edu.my

[c]School of Information Technology, Monash University, 46150 Bandar Sunway, Malaysia
Ali.Safaa@monash.edu

[d]School of Information and Communication Technology, Griffith University, Nathan Campus, Brisbane QLD 4111, Australia
ali.mirjalili@gmail.com

Abstract. Prediction of stock price is one of the most challenging tasks due to nonlinear nature of the stock data. Though numerous attempts have been made to predict the stock price by applying various techniques, yet the predicted price is not always accurate and even the error rate is high to some extent. Consequently, this paper endeavours to determine an efficient stock prediction strategy by implementing a combinatorial method of Grey Wolf Optimizer (GWO), Clustering and Non Linear Autoregressive Exogenous (NARX) Technique. The study uses stock data from prominent stock market i.e. New York Stock Exchange (NYSE), NASDAQ and emerging stock market i.e. Malaysian Stock Market (Bursa Malaysia), Dhaka Stock Exchange (DSE). It applies K-means clustering algorithm to determine the most promising cluster, then MGWO is used to determine the classification rate and finally the stock price is predicted by applying NARX neural network algorithm. The prediction performance gained through experimentation is compared and assessed to guide the investors in making investment decision. The result through this technique is indeed promising as it has shown almost precise prediction and improved error rate. We have applied the hybrid Clustering-GWO-NARX neural network technique in predicting stock price. We intend to work with the effect of various factors in stock price movement and selection of parameters. We will further investigate the influence of company news either positive or negative in stock price movement. We would be also interested to predict the Stock indices.

Keywords: **Stock prediction, Grey Wolf Optimizer, Non Linear Autoregressive Exogenous algorithm, K-means clustering, Error Rate.**

1. Introduction

For few decades, stock market is one of the best investment options for investors despite stock price being uncertain and unpredictable. Due to the nonlinear nature of stock, the prediction of stock price is even more complicated task. Moreover, the stock price of a company does not depend on the economic condition of a country. Researchers are continuously striving to find the best prediction technique for stock market as the economy of a country heavily depends on stock market.

The advancement of Information and Communication technology has made the processing of stock data much faster and the price of stock changes quite frequently. Due to this, the stock brokers, big investors like financial institutions, banks and any other investor need to make the stock investment decision within shortest possible time. However, the changes of stock prices are not depend on only one factor rather it is controlled by various factors like positive or negative news related to the company, political turbulences, rate of interest, dollar price, natural disasters and many more [1].

The investment decision to the stock market is a crucial task. The investor needs to be well educated and trained about the investment decision as the price of stock may be increased or decreased next day. Nevertheless, neural network has proved its potential as it can make better prediction than Fundamental and Technical analysis [2]. This paper is an attempt to guide the investors about the stock investment decision.

Data mining is a very important research dimension which is a way of discovering useful patterns from huge collection of data [3]. Large amount of historical data can be managed well through data mining technique as it is a way of converting data into more manageable form [4]. Clustering is a data mining method which can classify the unclassified data into smaller groups that has common properties [5]. Hence, the clustering data mining technique can be used for classifying the stock data.

Neural network is another very important method that consists of a lot of interconnected processors and it can have the capacity to deal with non-linear data like stock data which changes very fast [6]. Neural network is found to be very effective by several researches compared with traditional methods. It can be used for prediction of many different types of data including financial field i.e. prediction of stock and performance [7,8]. Classification is a data analysis method of neural network by which it can predict categorical class labels. Grey Wolf Optimizer (GWO) is a classification algorithm that can be optimized for solving a challenging problem with unknown search spaces [9]. Grey Wolf Optimizer is a process through which the algorithm is optimized for dealing with stock data. A nonlinear autoregressive exogenous model (NARX) is a nonlinear autoregressive time series modeling which can predict the recent value of a time series based on past values of same series.

In this research, we have collected data from renowned stock market like New York Stock Exchange (NYSE), NASDAQ and rising stock market like Dhaka Stock Exchange (DSE). We have employed hybridized technique on the stock data called Clustering-GWO-NARX, where the Clustering is a process by which the most promising stock is categorized and in this research work we have employed k-means clustering, next the classification algorithm called Grey Wolf Optimizer (GWO) is applied to determine the classification rate, then Nonlinear Autoregressive Exogenous (NARX) algorithm is applied to predict the stock price.

The remaining part of the paper is organized as follows. Section 2 presents brief overview of related research. Section 3 describes the basic concepts of GWO. Section 4 illustrates the architecture of the proposed prediction system, while Section 5 explains the experimental results obtained. The paper is concluded in section 6.

2. Related Research Work

Numerous researches have been directed towards stock prediction based on not only fundamental or technical analysis but also through data mining and neural network based applications. However, each technique has its own pros and cons. Moreover, none of the techniques has proved to be capable in making 100% correct prediction. Due to this reason, the research is on-going for developing an accurate stock prediction system. As data mining and neural network based technique has proved to be one of the most accurate stock prediction technique, we are going to review previous studies on data mining and neural network strategies applied for stock prediction in the field of our research study.

2.1. General stock prediction systems and their limitation

Prediction of stock price is very popular in financial and academic arena. Fundamental method is widely used in analysing the stock market through Time series analysis. Numerous existing forecasting systems are available based on time series data and factors for stock prediction. Xu has attempted to predict weekly stock price by mingling the conventional time series analysis technique with information from the Google trend website and the Yahoo finance website. He had found that there is a correlation between the changes in weekly stock prices and the values of important news/events computed from the Google trend website. He also argued that his result is better than traditional time series analysis however he has also agreed that his study has many drawbacks as well. He has finally suggested that further research can be conducted to improve the stock prediction and more accurate algorithm can be used for such prediction [10]. Statistical analysis is another way of making stock prediction by applying regression, co-linearity, ARIMA and so on. But, main drawback of statistical model is that it can deal with linear data whereas stock data is non-linear in nature. Due to this reason, the statistical model is becoming obsolete.

2.2. Stock prediction using data mining: Issues and Challenges
Extracting knowledge from database is known as Data mining. Data mining is able to extract the appropriate data based on various techniques i.e. Classification, Prediction, Clustering, Association. Classification and Prediction is predictive model, whereas Clustering and Association is descriptive model. Unknown pattern of data can be discovered through data mining [11]. Kannan et al. has emphasized that the amount of data is growing day by day and hence we need to apply an appropriate technique to better manage the huge amount of data. Therefore, data mining is identified as very demanding and one of the best key research areas in data management. So, they have applied data mining technology and performed data analysis to predict whether the closing price of a stock will increase or decrease for a day better than level of significance. However, their algorithm was able to make 50% correct prediction and their algorithm was silent for few situations i.e. when to sell the stock4. Hence, better data mining algorithm is needed to be explored to make more accurate prediction.

2.3. Stock prediction systems using neural network and their limitation
Neural network is established based on artificial intelligence and it consists of interconnected processors like human brain. It uses mathematical model to make prediction. The most exciting feature of neural network is its ability to deal with non-linear or dynamic data. Due to this, it is extensively applied in dealing with financial data i.e. forecasting the prices of stock market. Neural network can make good prediction if it is configured well for a particular problem as it can learn based on patterns. However, the main challenge of neural network is optimizing number of neurons for a problem and selection of data [11]. Tomasz Kozdraj has attempted to apply neural network in predicting stock price for Warsaw Stock Exchange. He has considered stock prices of 30 companies and performed necessary research and analysis which provided him average level of mean absolute percentage error. The error value was acceptable for his study however he encountered some disturbances in forecasting fitness. He added that though the neural network is worthwhile in stock prediction, but there is no existence of universal forecasting method. The problem which lowered his result is due to various reasons such as one single outcome of a network to a specific internal decision is very difficult, noisy data, overfitting which affects generalization, choice of input series, network parameter setting etc. He concluded that complex network suffers from various difficulties such as data pre-processing, selection of architecture and parameterization. Finally, he resolved that neural network is having limitation and problems in stock prediction which needs further research [12].

2.4. Use of classification in stock prediction systems and their limitation
Analysing the data to determine the possible course of action is known as classification, for example: stock data is analysed to know whether it is feasible to invest or not. The dynamic and non-linear nature of stock data demands the application of sophisticated algorithm. However, the technological advance has made us available the computational intelligence by providing various tools and algorithms. Classification and various machine learning algorithms which use combination of classifiers are also gaining researchers' attention in dealing with the stock prediction. Among the various classification methods, some of the researchers use single classification for stock forecasting, whereas others use combination of multiple classifiers to improve the accuracy of prediction [13]. Teixeira et al. have

proposed of applying combination of technical analysis and nearest neighbour classification in predicting Brazilian stock market. They have compared their result with buy-and-hold strategy and found the combination of technical analysis and classification produced more profits. Eventually, their proposed method also reduced the risk. However, they have suggested using of better classification algorithm such as Support Vector machines, k-NN etc. to achieve more generalization. Additionally, they have also recommended applying classification technique in Asian market [14]. B. Narayanan et al. have attempted applying two classification methods AdaSVM and AdaNaive and compared their result with widespread classification methods i.e. Support Vector Machine (SVM) and Naïve Bayes which gave them more accuracy with significant reduce of classification error. They have also mentioned that change of technique and increase of input data may provide even more accurate result. Finally, they have agreed that continuous research is essential in developing an acceptable stock forecasting method [15].

2.5. Stock prediction systems using hybrid data mining and neural network and their limitation

Data mining is capable of describing cause and effect relation of information, whereas, much prediction problems can be solved using neural network. However, distinct application of neither data mining nor neural network is as effective as the combination of these two techniques15. Prior researches also support that hybridization of these two techniques provide more accurate result with less error rate. Tsai et al. have applied hybridization of artificial neural network (ANN) and decision tree (DT) data mining technique in creating stock price forecasting model for Taiwan stock market which provided them 77% accuracy which was higher than the single use of ANN or DT. However, they recommended performing further research by applying other techniques such as support vector machine, genetic algorithm etc. They have also suggested pre-processing data applying genetic algorithm to achieve better prediction accuracy. Finally, they have added that other industries can be merged with electronic for comparisons [16]. The database of shares was studied and prediction was made accordingly applying data mining and artificial intelligence by Navale et al. They have gained highest probability prediction for share in their study. However, they have emphasized the necessity of improving parameters in accurate prediction performance through data mining and artificial intelligence. They have also recommended for further research on hybridizing artificial intelligence and data mining to reveal the weaknesses of other researches [17].

The advantages and shortcomings of the relevant researches above states that further research is required for revealing the uncertainty of stock market. In search of appropriate stock prediction method, most effective methods like data mining, neural network, classification etc. can be applied for developing accurate prediction system.

3. Basic concepts of Grey Wolf Optimizer (GWO)

An algorithm which is executed iteratively to produce optimum solution through the comparison of various solutions is known as Optimization algorithm. Grey Wolf Optimizer (GWO) is a new meta-heuristic algorithm which is emulated from grey wolves (Canis lupus) and it mimics the leadership hierarchy and hunting mechanism of grey wolves in nature[9]. The leadership hierarchy is simulated applying four types of grey wolves i.e. α, β, δ, ω and hunting is implemented applying three main steps i.e. searching for prey, encircling prey, attacking prey. The GWO algorithm is applicable to challenging problems in unknown search spaces which produces better result than PSO, GSA, DE, EP and ES9. In this research, GWO algorithm is analyzed and it is applied as a training algorithm for Multi-layer perceptron such that it suits with stock data. The novelty of the research is that GWO is applied first time for stock prediction. The proposed algorithm improves the wolves attack strategy. It calculates the weights based on wolves fitness function and gives the highest weight to the dominant wolf concurrently to improve the convergence. The empirical results confirm the power of GWO and demonstrate that the algorithm can faster decide the suitable thresholds, provide good classification rate, efficiency and accuracy.

Grey wolves (Canis lupus) belong to Canidae family and they are considered as apex predators, meaning that they are at the top of the food chain. They have a very strict social dominant hierarchy.

The algorithm divides the wolves into four types: α, β, δ, and ω, whereas, each type of wolves displays the following social behavior.

The leaders are a male and a female, called alpha. Decision making about hunting, sleeping place, time to wake etc. is made by alpha, that's the reason alpha becomes the leader in the pack and others follow its orders. The best solution of a problem can be determined identifying the location of alpha, as it is the best member in managing the pack.
The second level in the hierarchy of grey wolves is beta. The betas are subordinate wolves that help the alpha in decision-making or other pack activities. The beta maintains the discipline of the pack that enforces the alpha's commands in the pack and hence, it takes the role of the advisor to the alpha.
The third level in the hierarchy of grey wolves is delta and they need to submit to alphas and betas, but they dominate the omega. They are responsible for watching the boundaries of the territory, warning the pack in case of any danger, protecting and guaranteeing the safety of the pack, helping the alphas and betas when hunting prey, and providing food for the pack and caring for the weak, ill, and wounded wolves in the pack.
The last grey wolf in the hierarchy is omega and it may not be an important individual in the pack, but it has been observed that the whole pack face internal fighting and problems in case of losing the omega, which is harmful to the group structure.

Group hunting is another interesting social behavior of grey wolves in addition to the social hierarchy. The main phases of grey wolf hunting are as follows: searching for the prey; tracking, chasing, and approaching the prey; pursuing, encircling, and harassing the prey until it stops moving; attacking toward the prey.
The fittest solution is considered as the alpha (α) in designing mathematical model of the social hierarchy of wolves for designing GWO[9]. Beta (β) and delta (δ) are respectively the second and third fittest solution. Omega (ω) is considered for the rest of the candidate solutions. In the GWO algorithm, the hunting (optimization) is guided by α, β, and δ. The ω wolves follow these three wolves.

3.1 Preliminary of GWO Mathematical Model
The mathematical model of MGWO can be formed as per the equations below:

$$\vec{P} = \left| \vec{N}.\vec{Z}_a(i) - \vec{Z}(i) \right| \quad (1)$$
$$\vec{Z}(i+1) = \vec{Z}_a(i) - \vec{M}.\vec{P} \quad (2)$$

Where, i denotes the current iteration, $\vec{M}$ and $\vec{N}$ are coefficient vectors, $\vec{Z}_a$ is the position vector of the prey, and $\vec{Z}$ denotes the position vector of a grey wolf.

The vectors $\vec{M}$ and $\vec{N}$ are calculated as follows:

$$\vec{M} = 2\vec{q}.\vec{s}_1 - \vec{q} \quad (3)$$

$$\vec{N} = 2.\vec{s}_2 \quad (4)$$

Where, components of $\vec{q}$ are linearly reduced from 2 to 0 over the number of iterations and and is used for controlling the trade-off between exploitation and exploration. The following equations will be employed for updating the value of variable:

$$\vec{q} = 2 - i\,(2 / X_i) \quad (5)$$

$$\vec{Z}(i+1) = (\vec{Z}_1 + \vec{Z}_2 + \vec{Z}_3)/3 \quad (6)$$

Where, X_i denotes the number of iterations, $\vec{s}_1$ and $\vec{s}_2$ are random vectors between [0, 1] which are employed to find the optimal solution. Appropriate idea about the potential location of prey can be availed by Alpha, Beta and Delta, where they help the Omega to follow the suitable positions. The values of $\vec{Z}_1$, $\vec{Z}_2$ and $\vec{Z}_3$ can be obtained through the equations below:

$$\vec{Z}_1 = \vec{Z}_\alpha - \vec{M}_1.\vec{P}_\alpha \quad (7)$$

$$\vec{Z}_2 = \vec{Z}_\beta - \vec{M}_2.\vec{P}_\beta \quad (8)$$

$$\vec{Z}_3 = \vec{Z}_\delta - \vec{M}_3.\vec{P}_\delta \quad (9)$$

In iteration i, the best 3 solutions are respectively, $\vec{Z}_1$, $\vec{Z}_2$ and $\vec{Z}_3$. Where, the values of $\vec{P}_\alpha$, $\vec{P}_\beta$ and $\vec{P}_\delta$ are as below:

$$\vec{P}_\alpha = |\vec{N}_1.\vec{Z}_\alpha - \vec{Z}| \quad (10)$$

$$\vec{P}_\beta = |\vec{N}_2.\vec{Z}_\beta - \vec{Z}| \quad (11)$$

$$\vec{P}_\delta = |\vec{N}_3.\vec{Z}_\delta - \vec{Z}| \quad (12)$$

In this study, GWO is modified to suit it with stock data and train the MLP using GWO for achieving a precise neural network and for removing the drawback of the back-propagation algorithm. Here, two important steps are included for this method that comprises: firstly, train the neural network using GWO to determine the optimal initial weights and secondly, test the results of the proposed GWO approach. This approach can improve the efficiency of the back-propagation to seek global optima in the search space. For the proposed GWO approach, the weights are achieved as a vector of variables.

For this approach, Root Mean Square Error (RMSE) is the cost function which can determine the error between the actual value and predicted value which can be expressed as below:

$$\text{RMSE} = \sqrt{\sum_{i=1}^{n}(a_i - p_i)^2} \quad (13)$$

Where, n is the number of observations, a_i is the number of actual values and p_i is the number of predicted values from neural network. The lower RMSE value is expected for determining acceptable prediction and making the model acceptable.

3.2 Pseudocode of GWO

The steps used in MGWO are as follows:

Step 1: Initialize the grey wolf population, P_i (i = 1,2,3,.....n)

Step 2: Initialize q, M and N

Step 3: Calcualte the fitness of each search agent, where, $Z_\alpha, Z_\beta, Z_\delta$ are the best, second best and third best search agent consecutively

Step 4: While i<MaxIterations
update the position of current search agent for each search agent by equation 6
Update q, M and N

Calculate the fitness of all search agents
Update Z_α, Z_β, and Z_δ
i = i + 1
End While

Step 5: Return Z_α

Steps of MGWO algorithm are as follows:

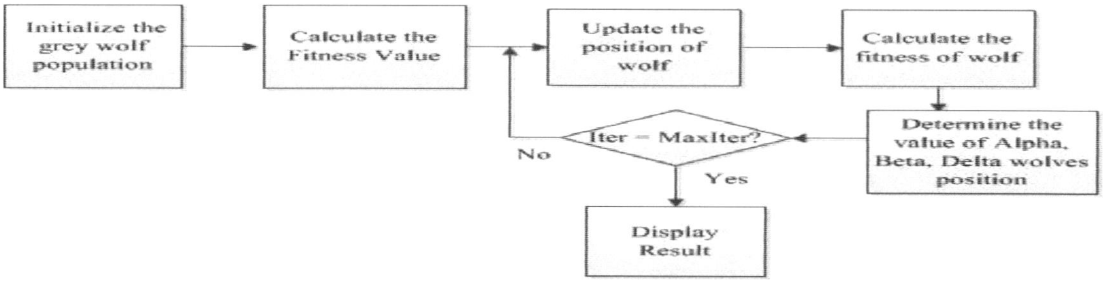

Figure 1. Steps of GWO Algorithm

4. Architecture of the proposed System

Literature has supported the application of data mining, classification and neural network technique for the prediction of non-linear stock market. In this study, we have collected data from New York Stock Exchange (NYSE) and NASDAQ through Yahoo Finance and from Dhaka Stock Exchange (DSE) through DSE Library/Website. We have collected about 6 years of stock data from 2011-2016. Kumar et al. have used five input parameters in predicting stock price applying Multi-Layer Perceptron with Back propagation neural network learning algorithm [18]. Many other prior researches also support us to consider the following parameter in stock prediction:

Open price, Close Price, Low Price, High Price, and Volume

We have pre-processed the data and then applied k-means clustering data mining technique to observe the growth of share price for each company. Consequently, k-means clustering helps us to identify the best company in terms of volume transacted, fluctuation of high price and low price of the shares. Classification algorithm GWO is applied on the selected company at this stage to determine the classification rate and decide whether the company can be selected for prediction or not. Then, we apply the optimized neural network Nonlinear Autoregressive Exogenous (NARX) is applied to predict the stock price, whether buy, sell or hold the shares. Finally, the prediction is evaluated by the cost function as explained in equation (13). We have also performed a comparative study of the prediction between prominent stock market and emerging stock market. The proposed system includes the following architecture:

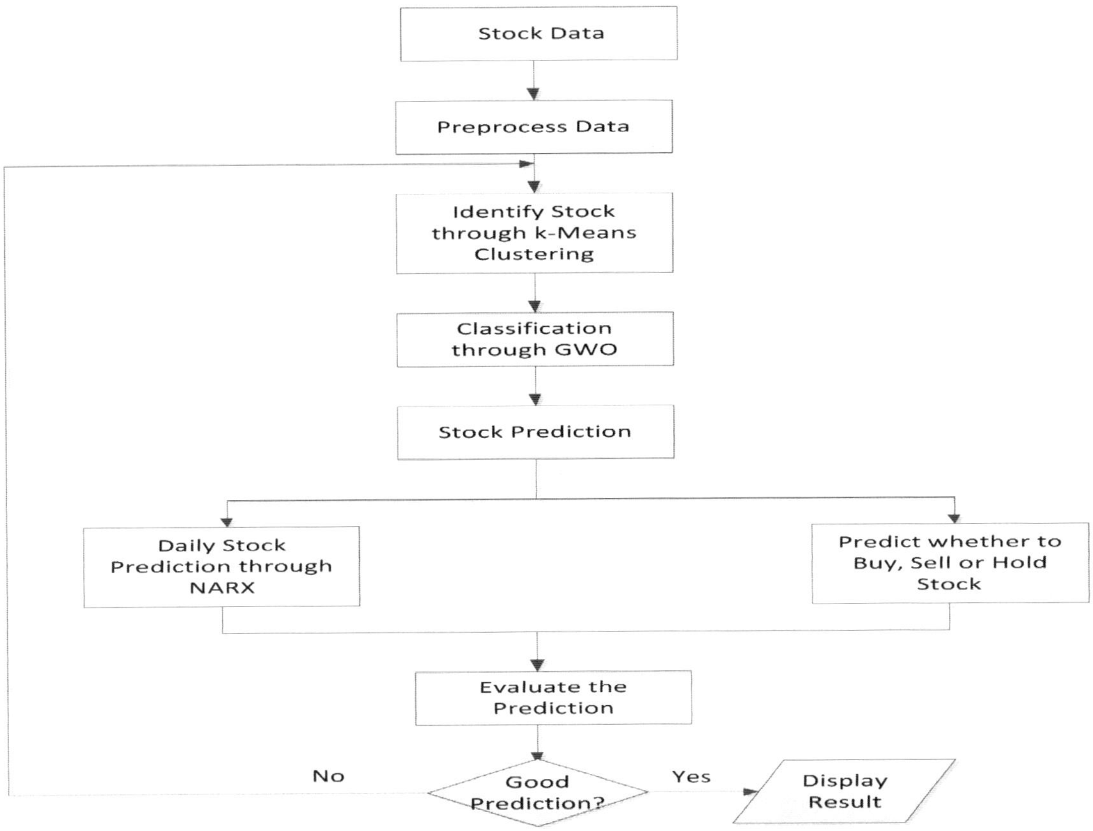

Figure 2. Architecture of the proposed prediction system

5. Experimental Results

The experiment for this work has been conducted in several stages. Firstly, the collected stock data is pre-processed and then clustering is applied to select the share to perform prediction. The raw stock data is pre-processed as per the Table 1 and Table 2 below to use it for clustering:

Table 1. Pre-processed stock data from NASDAQ for K-means clustering

Company Name	Trade Date	Low Price	High Price	Open Price	Closing Price	Volume of Trades
Microsoft	2014-1-2	37.1	37.4	37.35	37.16	30632200
Hudson Tech.	2014-1-2	3.59	3.74	3.70	3.69	192700
AMD	2014-1-2	3.84	3.98	3.85	3.95	2054800
NVIDIA Corp.	2014-1-2	15.72	15.98	15.92	15.86	6502300
Spartan Motors	2014-1-2	6.50	6.79	6.69	6.71	66400

Table 2. Pre-processed stock data from DSE for K-means clustering

Company Name	Trade Date	Low Price	High Price	Average Price	Closing Price	Volume of Trades
ACI	2013-02-12	140	147	145.07	145.04	8800
BXPHARMA	2013-02-12	56.70	65	62.49	62.51	568223
GLAXOSMITH	2013-02-12	547	547.10	547.01	547.01	218805
RECKITBEN	2013-02-12	725.00	725.00	725.00	725.00	500
RENATA	2013-02-12	788.20	798.00	793.71	793.85	4350
SQURPHARMA	2013-02-12	170.00	183.80	182.89	182.94	224,634

5.1 Clustering Result

K-means clustering places the data in two groups as we have selected two clusters to categorize the data based on number of volumes being traded. Clustering categorizes in a way that one cluster includes the pattern of data with highly growing, whereas another cluster contains the pattern with slowly growing. The following Table 3 lists the overall clustering data, though we will be using the highly growing data for classification.

Table 3. k-means Clustering Process

Attribute	Full Data (17092)	Cluster # 0 (12091)	Cluster # 1 (5001)
Company Name	ACI	ACI	SQURPHARMA
LOWPRC	1309.27	280.95	3795.47
HIPRC	1348.31	293.25	3899.15
AVGPRC	1330.63	288.11	3851.13
CLSPRC	1330.63	287.91	3851.63
TRDVOL	110681.94	147207.60	22373.25

The close observation of data pattern and also considering the growth, it has been determined that ACI has high growth and hence it has been selected for classification through GWO.

5.2 Classification Result through GWO

The Table 4 demonstrates the sample dataset used for classification, where 1 is used for Hold, 2 for Buy and 3 for Sell.

Table 4. Dataset used for Classification Process

Instrument	Low Price	High Price	Average Price	Closing Price	Volume	Decision
ACI, DSE	499	510.90	507.03	507.07	62446	1(Hold)
	490	510.00	504.89	504.76	61170	2(Buy)
	478	504.00	500.83	500.71	65774	2
	490	502.80	499.15	499.16	54554	1
	490	508.00	502.03	502.23	75680	1
	501	505.00	502.69	502.71	38720	1
	496.10	505.00	499.16	499.17	52370	1
	490.50	499.80	494.08	493.96	52000	2
	480	500.00	491.54	491.51	48281	1
	470	495.00	491.04	491.13	28930	2
	480	500.00	497.24	497.19	29702	3(Sell)

The classification through GWO produces the result indicated in Figure 3 and the classification rate through the algorithm is 100, which indicates that the proposed stock data is good for prediction [19, 20].

fx>> main

fobj =

@MLP_ACI

ans =

classification rate = 100

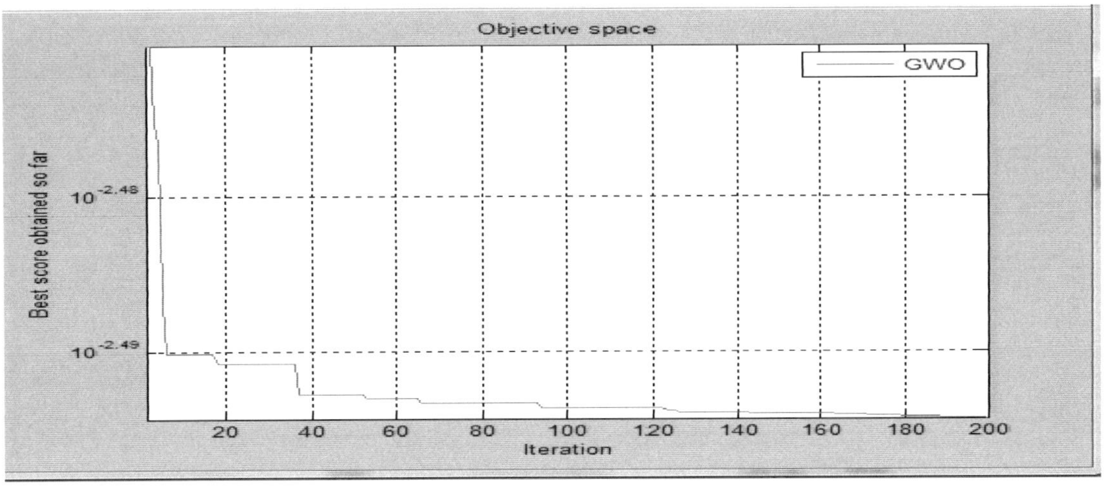

Figure 3. Result through MGWO

5.3 Result through Non Linear Autoregressive Exogenous (NARX) neural network model

The six years of daily stock price for companies from NYSE, Nasdaq, Bursa Malaysia and DSE has been provided to predict the High Price through Non Linear Autoregressive Exogenous neural network model which categorizes the data into Training, Validation and Test Set as indicated in Table 5. Where, Training Set consists 70% of total data, Validation Set contains 15% of data and Test Set holds 15% of data. The neural network model indicated in Figure 4 contains 10 Hidden Neurons and Number of Delays (d) is 2 in this architecture.

Table 5. Categorization of Data Sets

Dataset Type	Amount
Training Set	70% of the target timesteps
Validation	15% of the target timesteps
Testing	15% of the target timesteps

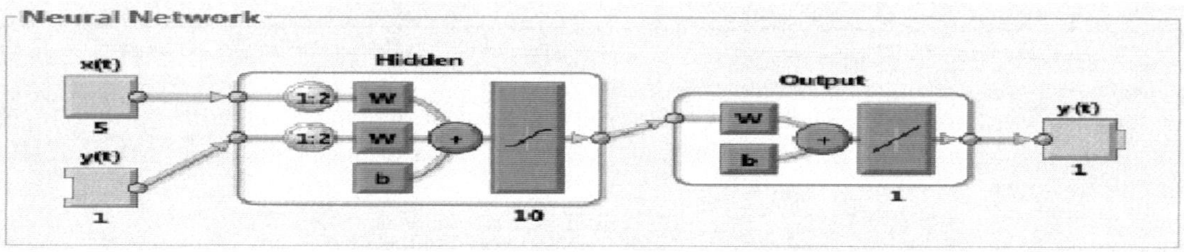

Figure 4. Architecture of the neural network Model

The equation 14 is applied to perform the stock prediction using Non-Linear Autoregressive Exogenous neural network.

$$P(k) = f(A(k-1),.....,A(d-k),P(k-1), ...,P(d-k)) \qquad (14)$$

Where, P(k) is the predicted Series, k is the time interval, d is the past values, P(k) is the actual price of the stock and A(k) is the target price to be predicted. The Table 6 lists Actual Value Vs Predicted value of stock price through Non Linear Autoregressive neural network algorithm for different stock market for a day, where the stock value for NYSE and Nasdaq in USD, stock value for Bursa Malaysia in Ringgit Malaysia (RM) and stock value for DSE in Bangladesh Taka (BDT).

Table 6. Actual Price Vs Predicted Price for various stock markets worldwide

Company Name	Stock Market	Date	Actual	Predicted	Forecasting Error (%)
Microsoft	Nasdaq,USA	2016-12-01	60.15	59.71	0.72
IBM	NYSE,USA	2016-12-01	162.2	160.966	0.76
Digi	Bursa Malaysia	2016-12-01	4.96	4.986	0.536
ACI	DSE, Bangladesh	2016-12-01	414	415.37	0.33

5.4 Performance Evaluation

The performance of the predicted price is evaluated through the equation 13 for various instruments of different stock markets produced the Root Mean Square Error (RMSE) value as indicated in Table 7 and the forecasting error related to the prediction is indicated in Figure 5. The RMSE value close to 0 indicates no error and prediction is completely acceptable. The forecasting error remains lower and reasonable through the proposed system. Hence, it can be concluded that the proposed hybridized system can be applied for stock prediction that can also reduce the error.

Table 7. RMSE value for instrument of various stock markets

Company Name	Stock Market	RMSE
Microsoft	Nasdaq,USA	0.7624
IBM	NYSE,USA	0.0199
Digi	Bursa Malaysia	0.0645
ACI	DSE, Bangladesh	0.2249

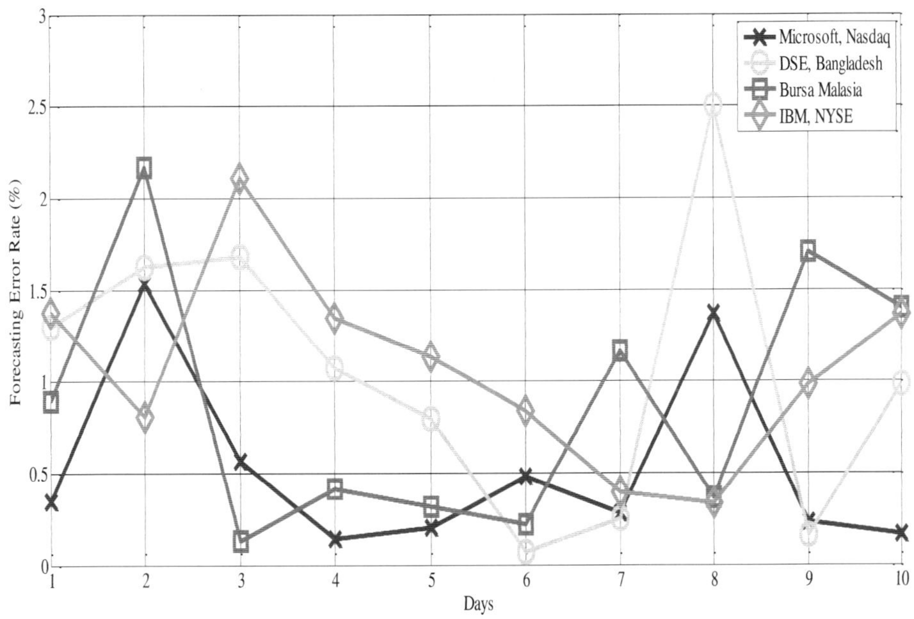

Figure 5. Comparison of Forecasting Error for various stock markets

The performance of a neural network can be also evaluated through generalization capability [21, 22]. The proposed prediction system can perform well for various stock markets and hence it has good generalization capacity.

6. Conclusion

This research attempts to construct a hybridized model for stock prediction and compare the performance of prediction between various stock markets. The empirical results through the hybridized model are satisfactory and the error rate also remained reasonable. Investors look for making good profit through stock investment and hence the proposed system can assist the investors in stock investment decision.

For some instrument the forecasting error remained high which is due to the moving of stock trend through various other factors. Our future plan is to work with the effect of various factors in stock price movement and selection of parameters. We would also take into consideration the influence of company news either positive or negative in stock price movement. Stock indices are another important component of stock prediction which can also be predicted.

7. References

[1] Bonde Ganesh and Khaled Rasheed 2012 *Proceedings of the International Conference on Genetic and Evolutionary Methods (GEM); Athens: 1-6. Athens: The Steering Committee of The World Congress in Computer Science, Computer Engineering and Applied Computing (WorldComp)*

[2] Coupelon O. 2007 Neural network modeling for stock movement prediction *Neural Network Modeling for Stock Movement Prediction*

[3] Jukic, N., Vrbsky, S. and Nestorov, S. 2014 *Database Systems: Introduction to Database and Data Wearhouses:* Pearson

[4] K. Senthamarai Kannan, P. Sailapathi Sekar, M.Mohamed Sathik and P. Arumugam 2010 *Proceedings of the International MultiConference of Engineers and Computer Scientists (IMECS); Hong Kong: 1-5, Financial Stock Market Forecast using Data Mining Techniques*

[5] Armano G. Marchesi M. and Murru A. 2005 A hybrid genetic-neural architecture for stock indexes forecasting *International journal of Information Science* Vol 170 pp 3-33

[6] Dase R.K. and Pawar D.D. 2010 Application of Artificial Neural Network for stock market predictions: A review of literature *International Journal of Machine Intelligence* ISSN: 0975–2927 Volume 2 Issue 2 pp 14-17

[7] Debashish Das, Ali Safa Sadiq and Noraziah Binti Ahmad 2016 An Efficient Time Series Analysis for Pharmaceutical Sector Stock Prediction by Applying Hybridization of Data Mining and Neural Network Technique *Indian Journal of Science and Technology* Vol. 9(21) pp 1-7

[8] Debashish Das, Ali Safa Sadiq, Noraziah Binti Ahmad and Jaime Lloret 2017 Stock Market Prediction with Big Data Through Hybridization of Data Mining and Optimized Neural Network Techniques *J. of Mult.-Valued Logic & Soft Computing* Vol 29 pp 157–181

[9] Seyedali Mirjalili, Seyed Mohammad Mirjalili and Andrew Lewis 2014 Grey Wolf Optimizer *Advances in Engineering Software* Vol 69 pp 46–61

[10] Selene Yue Xu 2014 Stock Price Forecasting Using Information from Yahoo Finance and Google Trend *Econ Honor Thesis* UC Berkeley

[11] Amit B. Suthar, Hiral R. Patel and Satyen M. Parikh 2012 A Comparative Study on Financial Stock Market Prediction Models *The International Journal of Engineering And Science (IJES)* Volume 1 Issue 2 pp 188-191

[12] Tomasz Kozdraj 2009 USING ARTIFICIAL NEURAL NETWORKS TO PREDICT STOCK PRICES *ACTA UNIVERSITATIS LODZIENSIS. FOLIA OECONOMICA* 225 pp 295-308

[13] Ballings, M., Van den Poel, D., Hespeels, N. and Gryp, R. 2015 Evaluating multiple classifiers for stock price direction prediction *Expert Systems with Applications* Vol 42 pp 7046-7056

[14] Lamartine Almeida Teixeira and Adriano Lorena Inácio de Oliveira 2009 A method for automatic stock trading combining technical analysis and nearest neighbor classification *Expert Systems with Applications* Issue 37 pp 6885–6890

[15] B. Narayanan and M. Govindarajan 2015 Prediction of Stock Market using Ensemble Model *International Journal of Computer Applications* Volume 128 No 1 pp 18-21

[16] Tsai, C.-F. and Wang, S.-P. 2009 Stock Price Forecasting by Hybrid Machine Learning Techniques *Proceedings of the International MultiConference of Engineers and Computer Scientists (IMECS)* Hong Kong

[17] G. S. Navale, Nishant Dudhwala, Kunal Jadhav, Pawan Gabda and Brij Kishor Vihangam 2016 Prediction of Stock Market using Data Mining and Artificial Intelligence *International Journal of Computer Applications* Volume 134 No 12 pp 9-11

[18] K.K.Suresh Kumar and N.M.Elango 2012 Performance analysis of stock prediction using ANN *Global journal of Computer Science and Technology* Volume 12 issue 1

[19] Seyedali Mirjalili and Ali Safa Sadiq 2011 Magnetic optimization algorithm for training multi layer perceptron *IEEE 3rd international conference on Communication software and networks (ICCSN)* 27-29 May Xi'an China pp 42-46

[20] Seyedali Mirjalili, Andrew Lewis and Ali Safa Sadiq 2014 Autonomous particles groups for particle swarm optimization *Arabian Journal for Science and Engineering* Vol 39 No 6 pp 4683-4697

[21] Kaastra, I. and Boyd, M.S 1996 Designing a Neural Network for Forecasting Financial and Economic Time Series *Neurocomputing* 10(3) pp 215-236

[22] MohdTahir Ismail, Buba Audu, Mohammed Musa Tumala and Edna Manga 2015 The Linear GARCH Modelling of Nigerian Stock Prices, *Journal of Computer Science & Computational Mathematics* Vol 5 pp 55-59

Acknowledgement

We would like to express our sincere gratitude and special thanks to everyone who has supported and made this paper possible. We would like to extend our acknowledgment of gratitude toward the significant contributors. Special thanks to Doctoral Scholarship Scheme (DSS), PGRS170332 authorities and University Malaysia Pahang (UMP) for supporting the research. We would like to thank Asia Pacific University of Technology and Innovation (APU) for most support and encouragement. We would also like to thank Bursa Malaysia Library and DSE library for providing the data. Finally, we sincerely thank to our parents, family, and friends, who provide the advice and support. The product of this paper would not be possible without all of their support and encouragement.

Eye Redness Image Processing Techniques

M. R. H. Mohd Adnan[1], Azlan Mohd Zain[1], Habibollah Haron[1], Razana Alwee[1], Mohd Zulfaezal Che Azemin[2], Ashraf Osman Ibrahim[3,4]

[1]Soft Computing Research Group, Faculty of Computing, UTM Johor Bahru, Malaysia
[2]Faculty of Mechanical Engineering, UTM Johor Bahru, Malaysia
[3]Faculty of Computer and Technology, Alzaiem Alazhari University, Khartoum, Sudan
[4]Arab Open University, Sudan

azlanmz@utm.my

Abstract. The use of photographs for the assessment of ocular conditions has been suggested to further standardize clinical procedures. The selection of the photographs to be used as scale reference images was subjective. Numerous methods have been proposed to assign eye redness scores by computational methods. Image analysis techniques have been investigated over the last 20 years in an attempt to forgo subjective grading scales. Image segmentation is one of the most important and challenging problems in image processing. This paper briefly outlines the comprehensive of image processing and the implementation of image segmentation in eye redness.

1. Introduction

Eye redness is a common sign of ocular disease which cause an excess of blood in the vessels in conjunctivae and sclera. The interpretation of eye redness is vary such as allergic conjunctivitis, morning eye congestion, dry eye, contact lenses complication, ocular drug-induced allergic reaction, marginal infiltrates of the cornea, bacterial conjunctivitis, severe ocular infection and iritis [1]. The increased diameter of blood vessels which is called as vasodilation are the factor that can cause an excess of blood vessels on the ocular surface. The important features that help the ophthalmologist to diagnose the eye redness are through the location of the vasodilation, the hue (depth of colour) and the intensity of the redness [1,2]. It is valuables in two areas to clinician to diagnose ocular surface characteristics earlier: firstly, it may provide additional diagnostic clues and second, refining the development of validated scales [1].

Image segmentation is one of the basic problems in image analysis. The task of image segmentation is to partition an image into a number of non-overlapping regions with homogeneous characteristics. A process of splitting the image into pixel bands is the image segmentation. As medical imaging contains uncertainties, there are difficulties in classification of images into homogeneous regions [3]. There is a need for segmentation algorithm for removing the noise from the medical image segmentation. Image segmentation is an important and challenging problem and a necessary first-step in image analysis as well as in high-level image interpretation and understanding such as robot vision, object recognition, and medical imaging. The main aim of the image segmentation is to segment the image into disjoint regions set with uniform and homogenous attributes such as tone, colour, texture, or intensity, etc., Still in many real situations for image segmentation issues such as poor contrast, noise and intensity in-homogeneities variation, limited spatial resolution, overlapping intensities make this segmentation as a difficult task [4].

This paper focuses on overview of image processing and its relation in segmenting eye redness. Previous studies show that image processing has been investigated in an attempt to forgo subjective grading scales. This paper outlines an understanding of how image processing operates to segmenting the eye redness. Since there are very limited numbers of studies on the image processing in eye redness, it can be considered as the contribution of this paper. The objective of this paper is to study the relation of image segmentation as the image analysis.

The remaining sections of this paper are organized as follows: In Section 2, briefly discussed about clinical judgement of eye redness. In Section 3, Digital image processing are described in details. Section 4, shows the image segmentation in eye redness has been applied. Application of image processing in ocular research is mentioned in Section 5. To conclude this paper, the conclusion is presented in Section 6.

2. Clinical Judgement of Eye Redness

The assessment of the ocular surface redness is a routine procedure in clinical practice. The most commonly used method is subjective observation and description [5]. Clinical grading may be judged using at least two general strategies. The first is primarily luminance–chromaticity based. Judgments are made on the basis of the overall redness and brightness (luminance) of the eye. As the redness increases, so the luminance decreases. A second strategy is made on the basis of the appearance of the visible vessels. This could include judgments of the diameter of vessels, vessel tortuosity, and the proportion or number of vessels occupying the area to be graded. The difference between these methods is really one of scale: Luminance judgments would correlate with vessel appearance if the capillary beds giving rise to the conjunctival flush were resolved. Similarly, with a sufficiently low resolution, smaller vessels would not be resolved and would "blend" into the background redness. For any typical clinical observation, however, each type of judgment is possible and could vary (to a large extent) independently of the other. The automated and objective approach proposed is based on the same two criteria: Two features are extracted from each image, one based on redness and the other based on the appearance of blood vessels [6].

The clinical judgment of ocular redness is complex and poorly understood. Typically, the appearance of the eye is judged based on a scale, and the examination of these scales provides a lesson in contemporary views of measurement. Theoretical examination aside, the scales themselves are typically poorly described with few exceptions have been untested. Lack of understanding of the scales themselves, there is no empirical information about how clinicians make judgments of redness [7]. Thus, highly reproducible and fully automated methods of analysis would clearly be of great benefit to many medical staff and patients.

Clinical evaluation of conjunctival hyperemia is subjective, and relies on grading scales such as the McMonnies/Chapman-Davies scale [8], the Institute for Eye Research scale, the Efron scale [9], and a validated bulbar redness scale [10]. The grades generally range from zero to four or five. The scoring scales in these systems increment in single units. A clinician grading an eye with bulbar hyperemia halfway between grade 0 and grade 1 on the Efron scale will give a grade of 0.5. Thus, a weakness of this method is that it cannot provide a continuous linear quantitative evaluation.

In the past 10 years, numerous methods have been proposed to assign redness scores by computational methods. The question then might be, how does the procedure developed here advance these methods? There are a number of reasons why this experiment describes methods and results that actually make it feasible to use this technique as a replacement for grading the redness of images. First, the technology is readily available and expensive. Whereas some previous studies have used rather exotic hardware and software combinations, the algorithms are available to anyone using a computer running almost every operating system that may be encountered. Second, a minimum amount of operator intervention is required. This removes some of the subjectivity in some previous techniques and further lends itself to automation. And finally, it has shown that both accuracy and much less variability are present in the automated technique than in the subjective technique [11,12]. These factors, provide a strong rationale for the adoption of emerging image processing technique to replace clinical grading of bulbar redness. Because anterior segment assessment is much more than just redness evaluation, how to implement this technique more generally to replace in vivo grading is yet to be determined [13,14].

There is an advanced technology called the OCULUS Keratograph® 5M. It helps Ophthalmologist and Optometrists in accurately diagnosing eye redness and documenting the findings. The Keratograph® 5M can classify the eye and limbal redness using the R-Scan. The R-Scan detects vessels in the conjunctiva and evaluates the degree of redness. This eliminates the need for time-consuming comparisons and provides greater reliability during evaluation. Figure 1 shows the OCULUS Keratograph® 5M machine.

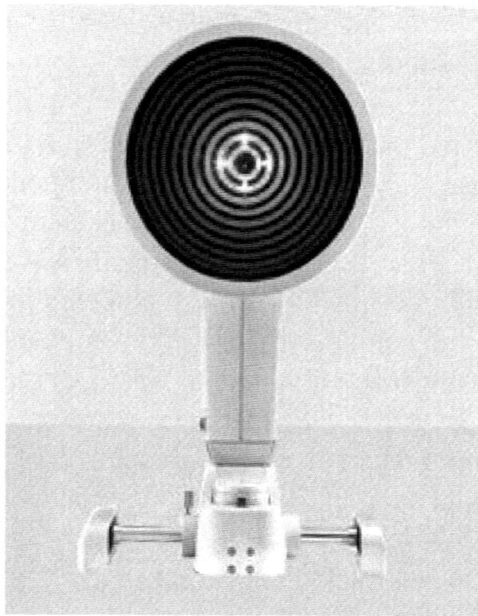

Figure 1. OCULUS Keratograph® 5M

R-scan software is automatic prediction of the eye redness. Conjunctival redness used to be an assessed subjectively and depended on the examiner. The R-Scan is the first module that automatically and objectively documents and classifies the eye and limbal degree of redness. The R-Scan detects the blood vessels in the conjunctiva and evaluates the degree of redness. This automatic assessment reduces time consuming and expertise of having to make manual comparisons using classification sheets.

3. Digital Image Processing

Digital image processing and image analysis have been used to derive physical attributes that describe bulbar redness chromatically and spatially. The Fundamental step in Digital Image processing is shown in Figure 2. The goal of image processing is to transform or enhance an image by Owen et al [15]:-

IOP Conf. Series: Journal of Physics: Conf. Series **892** (2017) 012019 doi:10.1088/1742-6596/892/1/012019

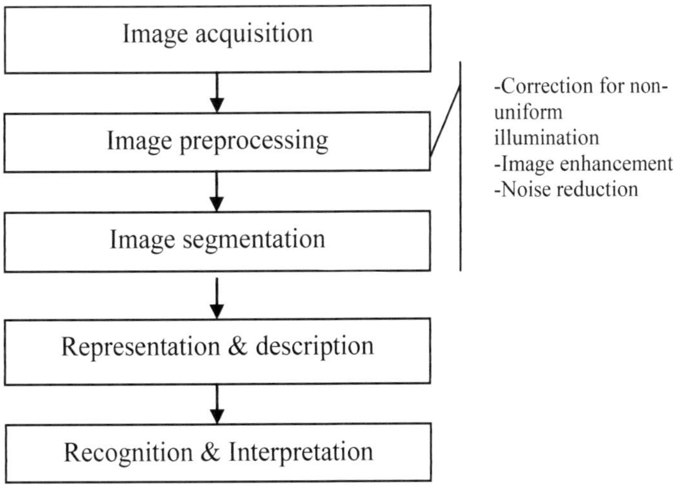

Figure 2. Fundamental step in Digital Image Processing

i)Image acquisition

- Acquire a digital image.

ii) Image pre-processing

- To improve the low image quality due to low dose usage [23]. Image enhancement techniques give a better result of the original image which is often presented as dark, low in contrast and noisy [24].

iii) Image segmentation

- To partitions an input image into its constituent objects.

iv) Representation and description (Feature selection and Feature extractions)

- Representation: To convert the input data to a form suitable for computer processing.
- Description: To extract features that results in some qualitative information of features that are basic for differentiating one class of objects from another.

v) Recognition and interpretation

- Recognition: To assign a label to an object based on the information provided by its descriptors.
- Interpretation: To assign meaning to an ensemble of recognized objects.

4. Image Segmentation

Image segmentation is performed by a set of techniques and its aim is to divide the image into a shape that is more conducive for analysis. Any technique should preserve the essential properties of the images; otherwise, there could be serious errors of analysis [13]. Image analysis techniques have been investigated over the last 20 years in an attempt to forgo subjective grading scales. Thresholding, edge detection, smoothing, color extraction, morphometry, and densiometry have all been used to measure eye redness [5]. We highlighted the two most important of image segmentation used in segmenting eye redness.

i) Threshold

Thresholding is the easiest way of segmentation. Segmentation through thresholding has fewer computations compared to other techniques [8].**Threshold segmentation** techniques grouped in classes [4]:

- **Local** techniques are based on the local properties of the pixels and their neighbourhoods.
- **Global** techniques segment an image on the basis of information obtain globally (e.g. by using image histogram; global texture properties).
- **Split, merge and growing** techniques use both the notions of homogeneity and geometrical proximity in order to obtain good segmentation results [8].

ii) Edge based

In this edge based segmentation, there is no need for the detected edges to be closed. Edges are detected to identify the discontinuities in the image. Edges on the region are traced by identifying the pixel value and it is compared with the neighbouring pixels. There are various edge detectors that are used to segment the image such as canny detection, roberts detection, prewitt detection, and sobel detection.

Edge detection contains Filtering, Enhancement and Detection [26]:

- **Filtering** is used to eliminate the images that often corrupted by unnecessary information which is called as noise.
- **Enhancement** it is essential to determine changes in intensities in the neighbourhood pixel.
- **Detection** determines which points of pixel are edge points because it may vary each particular point in each image.

5. Application in Ocular Research

Image processing and analysis have been used for the objective quantification of redness. In general, these studies analyzed sample images using image processing techniques, and compared the physical attributes to subjective estimates using grading. The image processing operations that were commonly applied to enhance the photographs of eye redness aimed at the removal of noise from the images and at separating the vessels from the background so that spatial attributes of the vessels could be quantified [16,17]. In general, noise reduction was achieved by filtering operations that are typically applied locally. The advantage of using localized filtering operations is that small capillaries, which may have very similar grey level intensities as the scleral background, are not removed from the image [18,19] investigated whether subjective grading estimates obtained using the Efron and IER grading scales could be predicted by image analysis techniques (edge detection and relative colour extraction). In agreement with [21] they found that a combination of morphometric and colour-based information was best suitable to predict subjective redness estimates.

Wang [20] identified surrounding ocular surface vessels in four zones around the cornea and directly calculate the vessel parameters to assess the degree of hyperaemia. They found a new method that directly calculated the parameters of ocular surface vessels around the limbus to assess the degree of regional bulbar redness. It offers advantages for objective observation of ocular surface redness, especially for the follow-up of disease treatment in individual patients.

Sanchez [14] proposed an automatic methodology to measure the redness level of the conjunctiva. The methodology proposed gave a good agreement for the transformation from the extracted features to grading scales, using artificial neural networks.

Park [22] established a new clinical grading scale and objective measurement for eye redness. The image segmentation which is using k-means and contrast-limited adaptive histogram equalization [CLAHE] algorithm showed the best correlation and reproducibility with subjective grading scale.

Downie [25] compared whether quantification of ocular eye redness using image processing of relative Red-channel (Red-value) can be applied to subjective grading scale and automated eye redness grading technique (Oculus Keratograph 5M, R-Scan). It revealed that Red-value calculations can be implemented to produce reliable estimates of ocular redness.

6. Conclusion

In this study, the image is segmented using a series of decision and there is no universal segmentation method for eye redness images and also an image can be segmented by using different segmentation methods. Objective image analysis of the anterior eye is applicable as being substantially more sensitive and reliable than subjective grading. It may therefore offer a new gold standard in the anterior ocular examination and could be developed further as a tool for use in research, to allow more highly powered analysis without bias, and in clinical practice to enhance the monitoring of anterior eye disease.

Acknowledgement

Special appreciative to reviewer(s) for the useful advices and comments. The authors greatly acknowledge the Research Management Centre, UTM for financial support through the Research University Grant (GUP) vot. no. Q.J130000.2528.16H81.

References

[1] Prabu C, Bavithiraja S V M G and Narayanamoorthy S 2016 A novel brain image segmentation using intuitionistic fuzzy C means algorithm. *International Journal of Imaging Systems and Technology*, 26(1), 24-28.

[2] Wang X Y, Wu Z F, Chen L, Zheng H L and Yang H Y 2016 Pixel classification based color image segmentation using quaternion exponent moments. *Neural Networks*, 74, 1-13.

[3] Abelson M B, Lane K and Maffei C 2010 Code red: The key features of hyperemia. *Rev Ophthalmol*, 17(4), 92-94.

[4] Zaitoun N M and Aqel M J 2015 Survey on image segmentation techniques. *Procedia Computer Science*, 65, 797-806.

[5] Rodriguez J D, Johnston P R, Ousler III G W, Smith L M and Abelson M B 2013 Automated grading system for evaluation of ocular redness associated with dry eye. *Clinical ophthalmology*, 7, 1197-1204.

[6] Sağ T and Çunkaş M 2015 Color image segmentation based on multiobjective artificial bee colony optimization. *Applied Soft Computing*, 34, 389-401.

[7] Zhao W J, Duan F, Li Z T, Yang H J, Huang Q and Wu K L 2014 Evaluation of regional bulbar redness using an image-based objective method. *International journal of ophthalmology*, 7(1), 71-76.

[8] Mcmonnies C W and Chapman-davies A 1987 Assessment of conjunctival hyperemia in contact lens wearers. Part II. *Optometry and Vision Science*, 64(4), 251-255.

[9] Efron N 1997 Clinical application of grading scales for contact lens complications. *Optician*, 213(5604), 26-34.

[10] Schulze M M, Jones D A and Simpson T L 2007 The development of validated bulbar redness grading scales. *Optometry & Vision Science*, 84(10), 976-983.

[11] Yogamangalam R and Karthikeyan B 2013 Segmentation techniques comparison in image processing. *International Journal of Engineering and Technology* (IJET), 5(1), 307-313.

[12] Peterson R C and Wolffsohn J S 2007 Sensitivity and reliability of objective image analysis

compared to subjective grading of bulbar hyperaemia. *British Journal of Ophthalmology*, 91(11), 1464-1466.

[13] Shahverdi R, Tavana M, Ebrahimnejad A, Zahedi K and Omranpour H 2016 An improved method for edge detection and image segmentation using fuzzy cellular automata. *Cybernetics and Systems*, 47(3), 161-179.

[14] Sánchez L, Barreira N, Pena-Verdeal H and Yebra-Pimentel E 2015 A novel framework for hyperemia grading based on artificial neural networks. In *International Work-Conference on Artificial Neural Networks*, 263-275.

[15] Owen C G, Ellis T J, Rudnicka A R and Woodward E G 2002 Optimal green (red- free) digital imaging of conjunctival vasculature. *Ophthalmic and Physiological Optics*, *22*(3), 234-243.

[16] Wu S, Hong J, Tian L, Cui X, Sun X and Xu J 2015 Assessment of bulbar redness with a newly developed keratograph. *Optometry & Vision Science*, 92(8), 892-899.

[17] Derakhshani R, Saripalle S K and Doynov P 2012 Computational methods for objective assessment of conjunctival vascularity. In *Engineering in Medicine and Biology Society* (EMBC),1490-1493.

[18] Ciesielski K C, Herman G T and Kong T Y 2016 General theory of fuzzy connectedness segmentations. *Journal of Mathematical Imaging and Vision*, 55(3), 304-342.

[19] Xu Z, Jiang H, Tao A, Wu S, Yan W, Yuan J, Liu C, DeBuc D C and Wang J 2015 Measurement variability of the bulbar conjunctival microvasculature in healthy subjects using functional slit lamp biomicroscopy (FSLB). *Microvascular research*, 101, 15-19.

[20] Wang X Y, Wu Z F, Chen L, Zheng H L and Yang H Y 2016 Pixel classification based color image segmentation using quaternion exponent moments. *Neural Networks*, 74, 1-13.

[21] Baudouin C, Barton K, Cucherat M and Traverso C 2015 The measurement of bulbar hyperemia: challenges and pitfalls. *European journal of ophthalmology*, 25(4), 273-279.

[22] Park I K, Chun Y S, Kim K G, Yang H K and Hwang J M 2013 New Clinical Grading Scales and Objective Measurement for Conjunctival Injection New Clinical Grading Scales. *Investigative ophthalmology & visual science*, 54(8), 5249-5257.

[23] Ahmad S A, Taib M N, Khalid N E A and Taib H 2012 An analysis of image enhancement techniques for dental X-ray image interpretation. *International Journal of Machine Learning and Computing*, 2(3), 292-297.

[24] Ahmad S A, Taib M N, Khalid N E A and Taib H 2012 Correlation between Quantitative and Qualitative Analysis on Image Quality of Digital Dental X-ray Images. *Journal of Computer Science & Computational Mathematics*, 2, 43-51.

[25] Downie L E, Keller P R and Vingrys A J 2016 Assessing ocular bulbar redness: a comparison of methods. *Ophthalmic and Physiological Optics*, 36(2), 132-139.

[26] Senthilkumaran N and Rajesh R 2009 Edge detection techniques for image segmentation–a survey of soft computing approaches. *International journal of recent trends in engineering*, 1(2), 250-254.